becoming

being

passing

becoming

being

passing

Sabine Pankert

verbavolant

Stages

As every flower fades and as all youth
Departs, so life at every stage,
So every virtue, so our grasp of truth,
Blooms in its day and may not last forever.
Since life may summon us at every age
Be ready, heart, for parting, new endeavor,
Be ready bravely and without remorse
To find new light that old ties cannot give.
In all beginnings dwells a magic force
For guarding us and helping us to live.

Serenely let us move to distant places
And let no sentiments of home detain us.
The Cosmic Spirit seeks not to restrain us
But lifts us stage by stage to wider spaces.
If we accept a home of our own making,
Familiar habit makes for indolence.
We must prepare for parting and leave-taking
Or else remain the slaves of permanence.

Even the hour of death may send
Us speeding on to fresh and newer spaces,
And life may summon us to newer races,
So be it, heart: bid farewell without end.

—Hermann Hesse

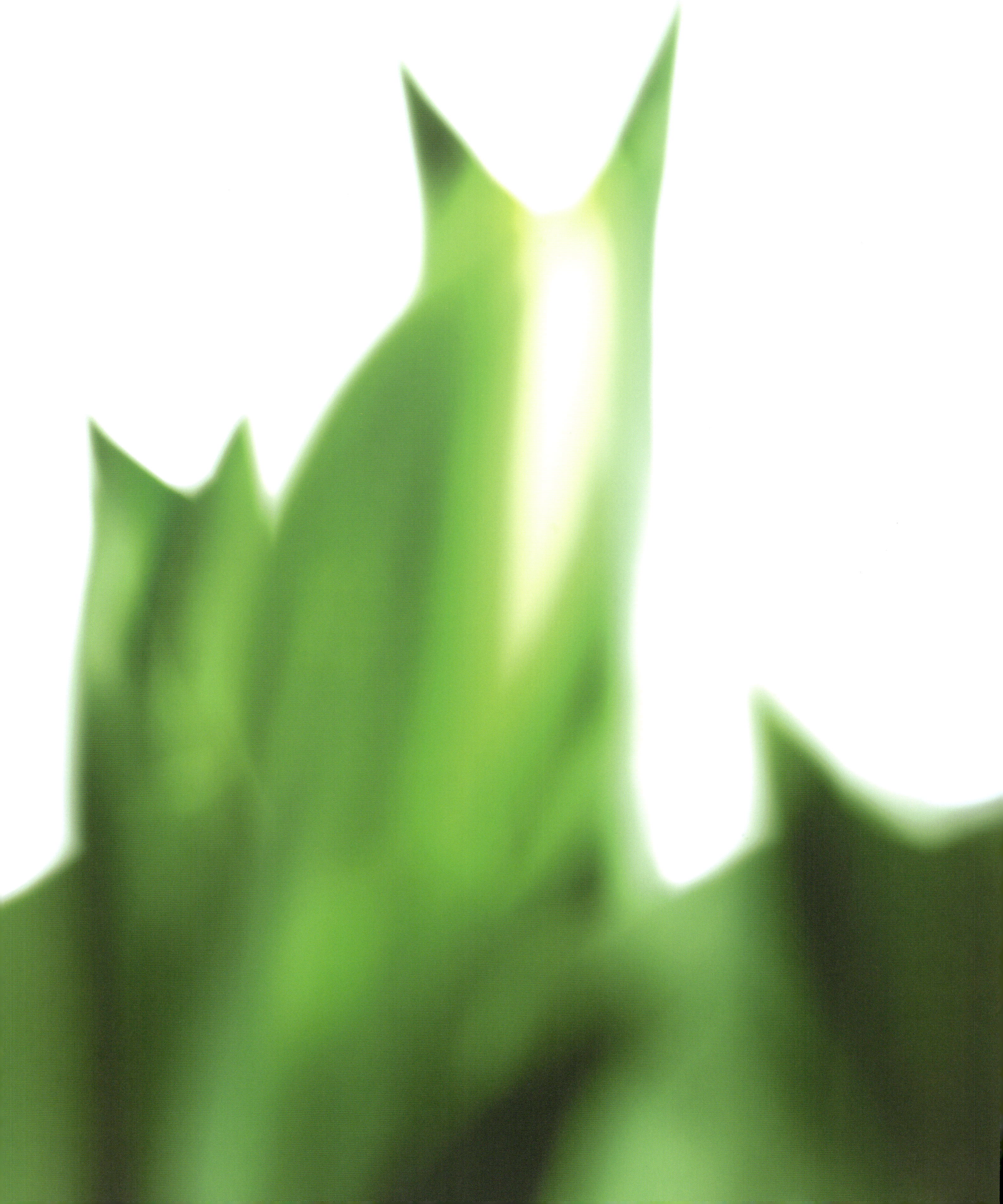

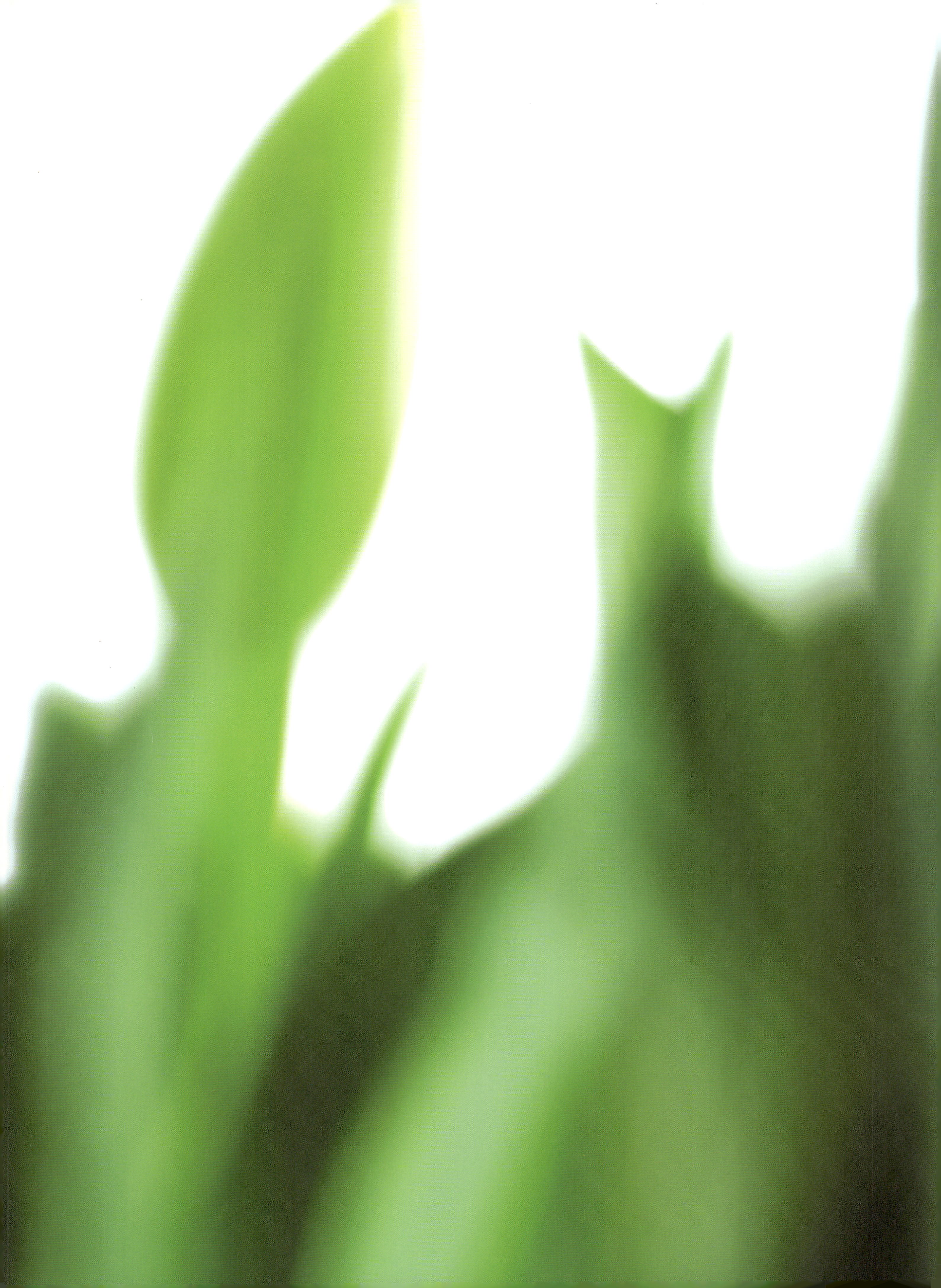

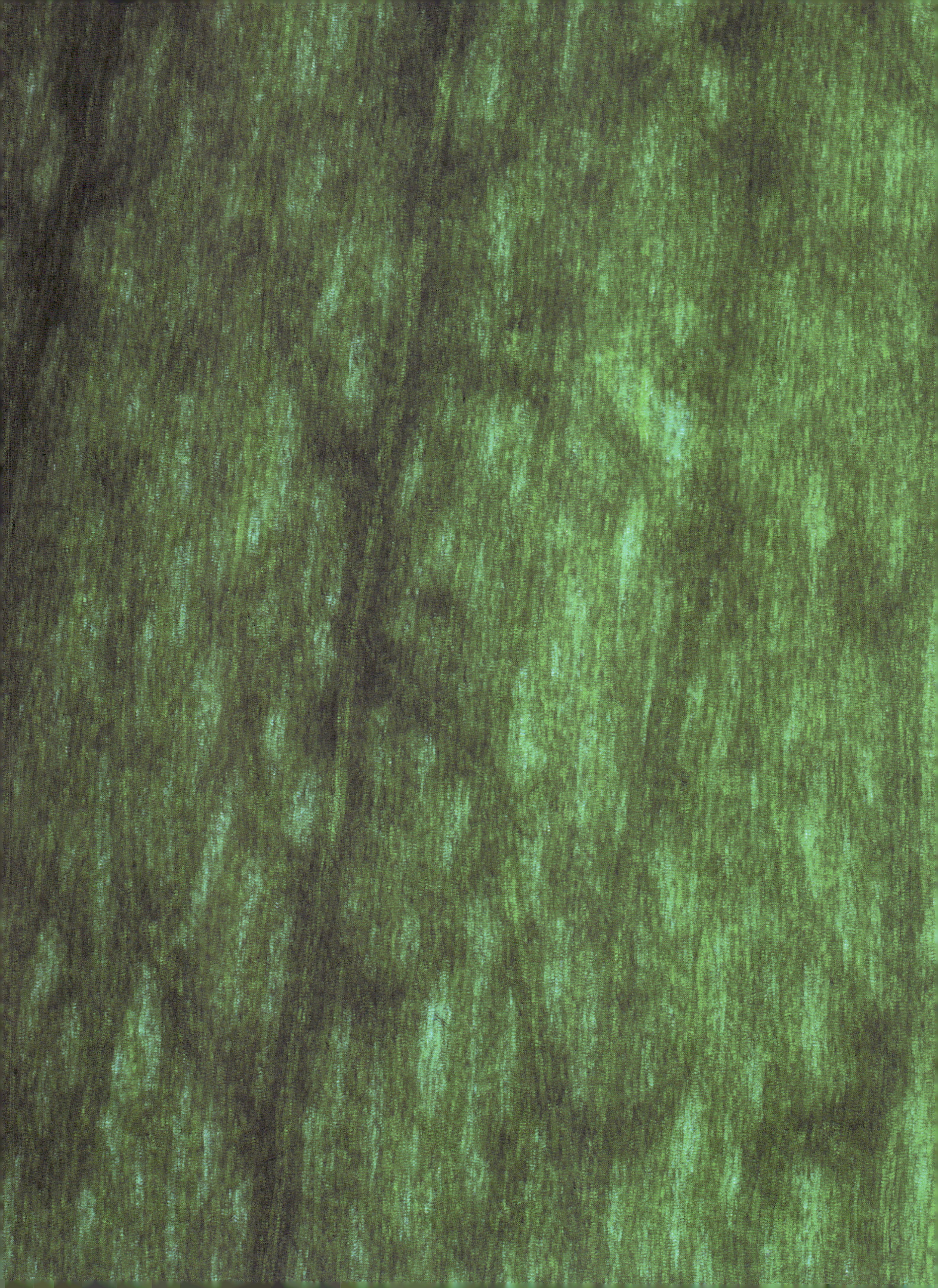

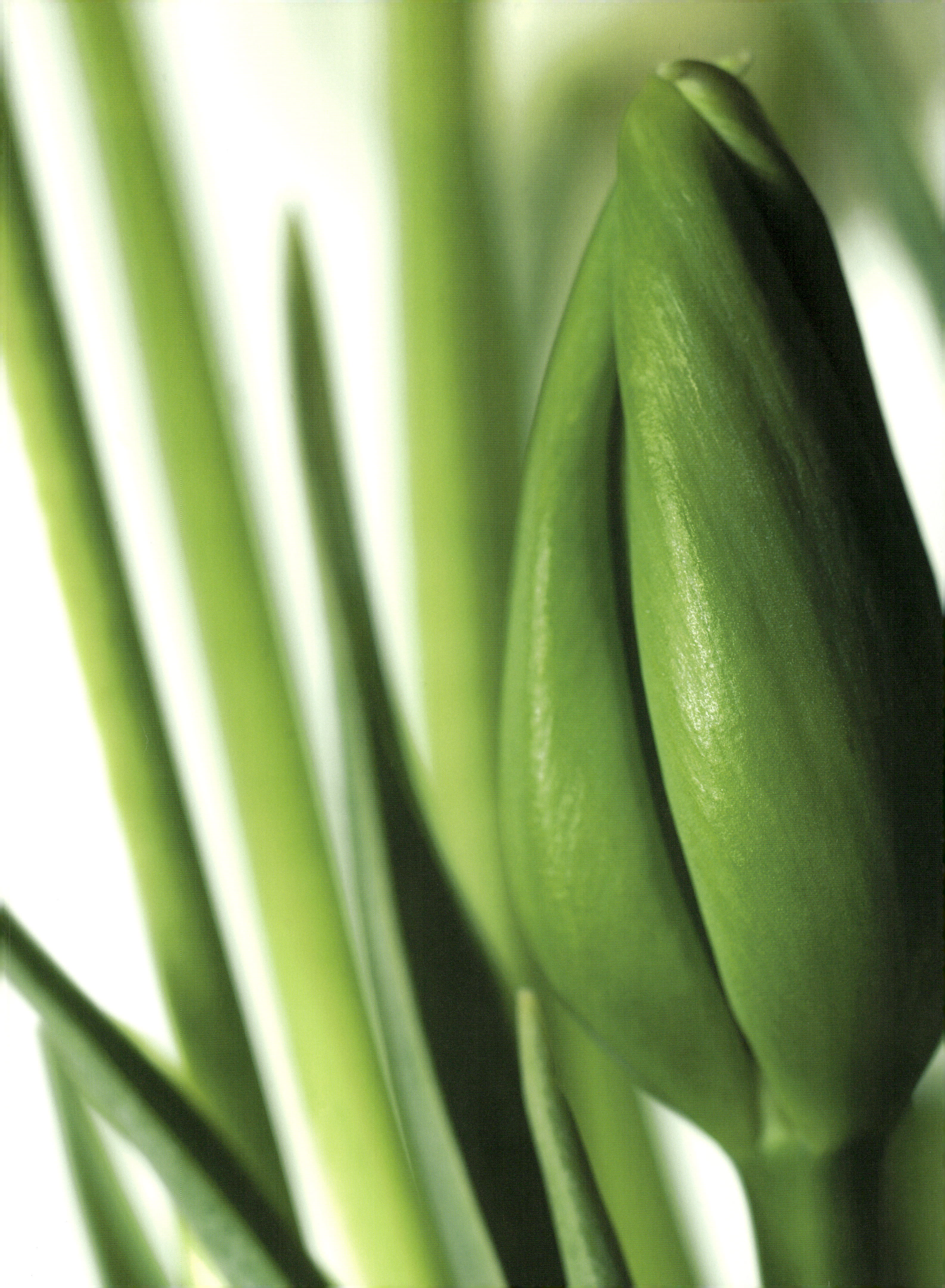

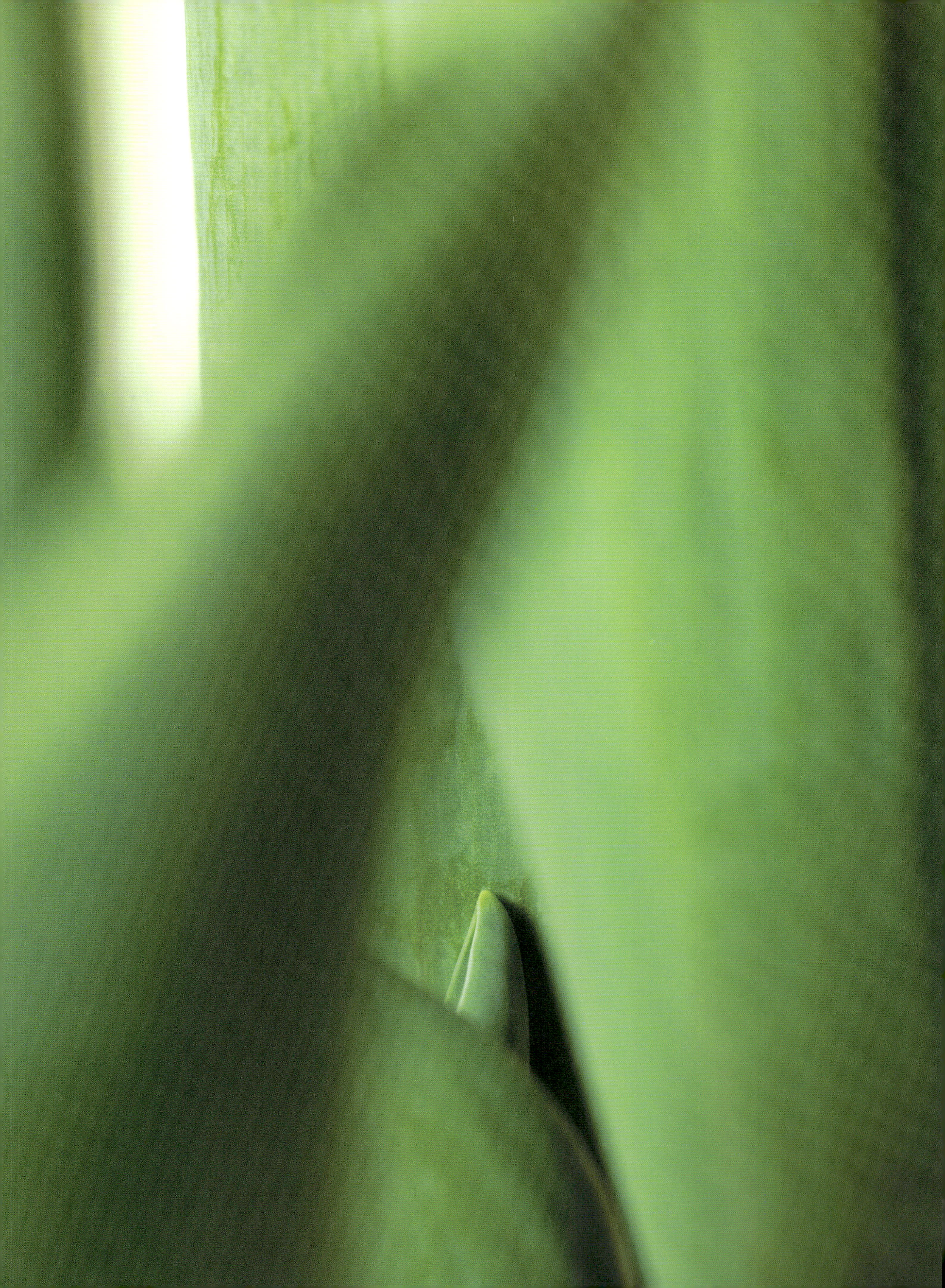

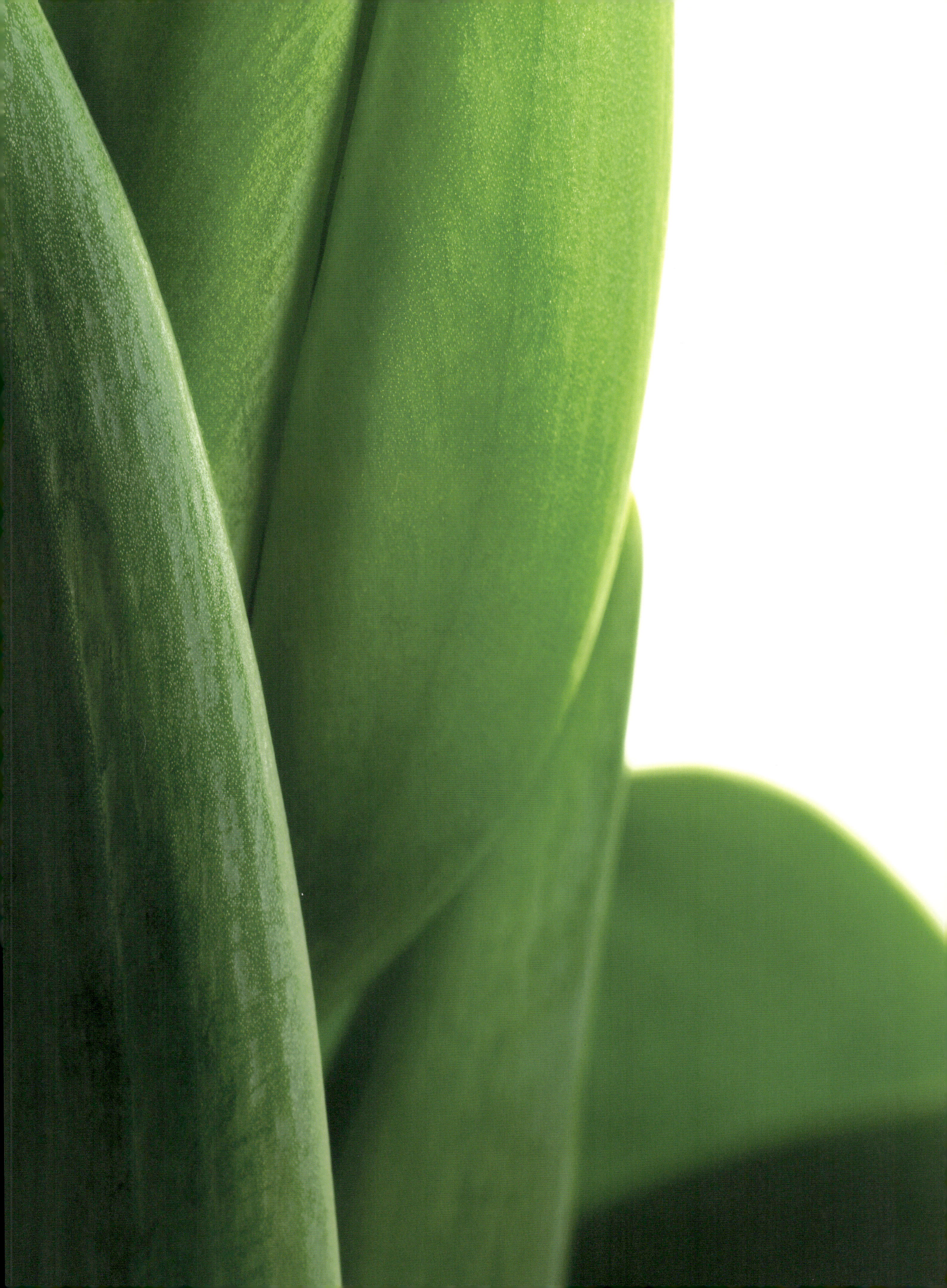

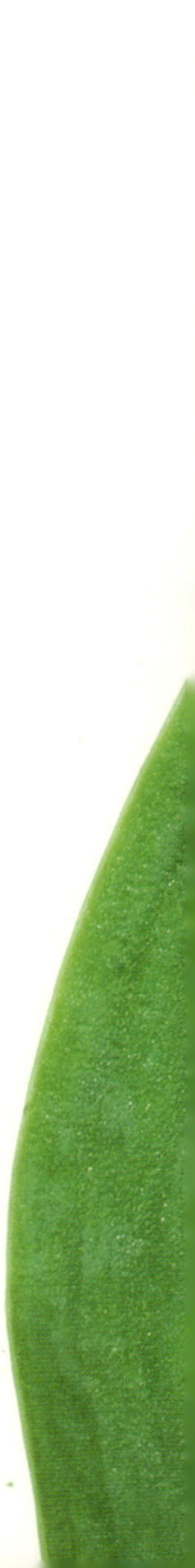

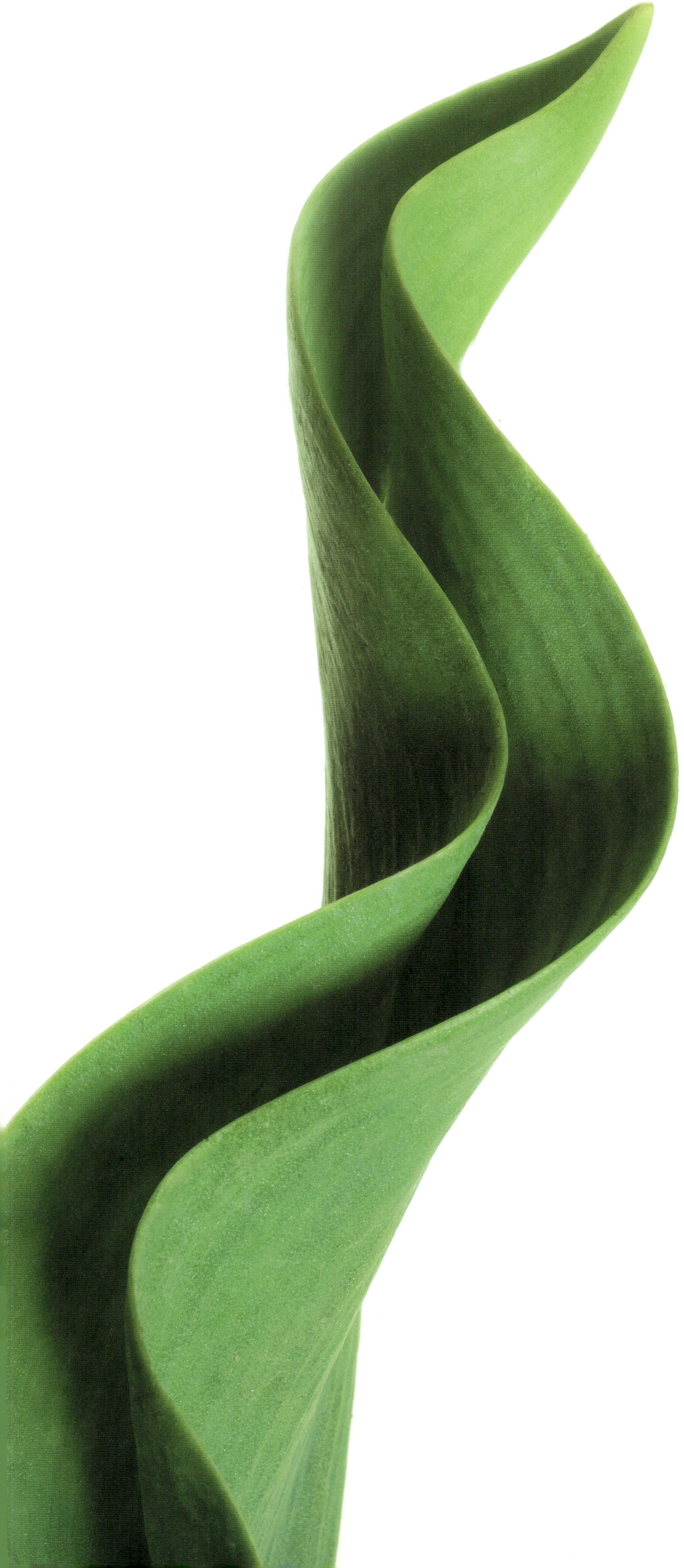

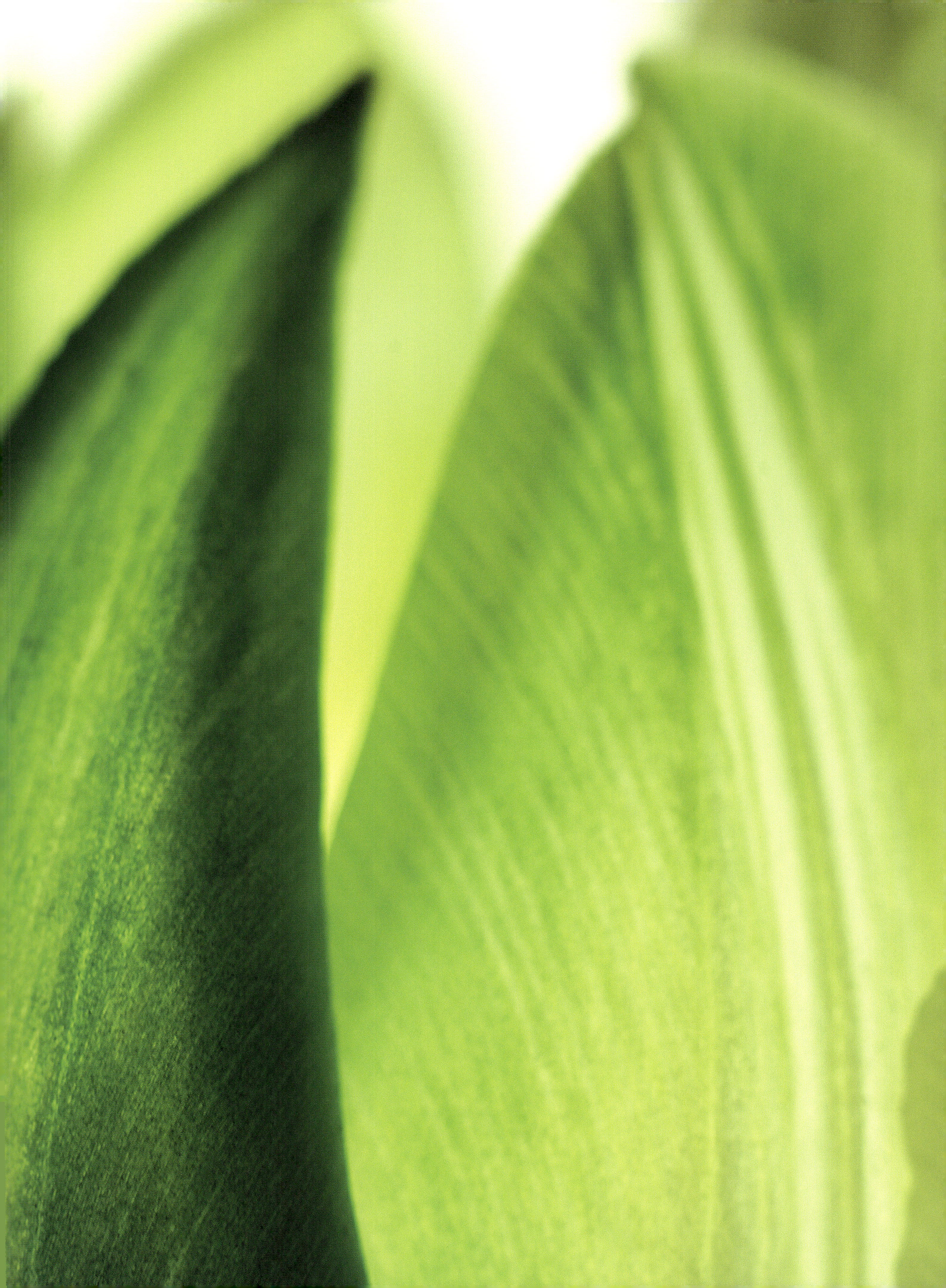

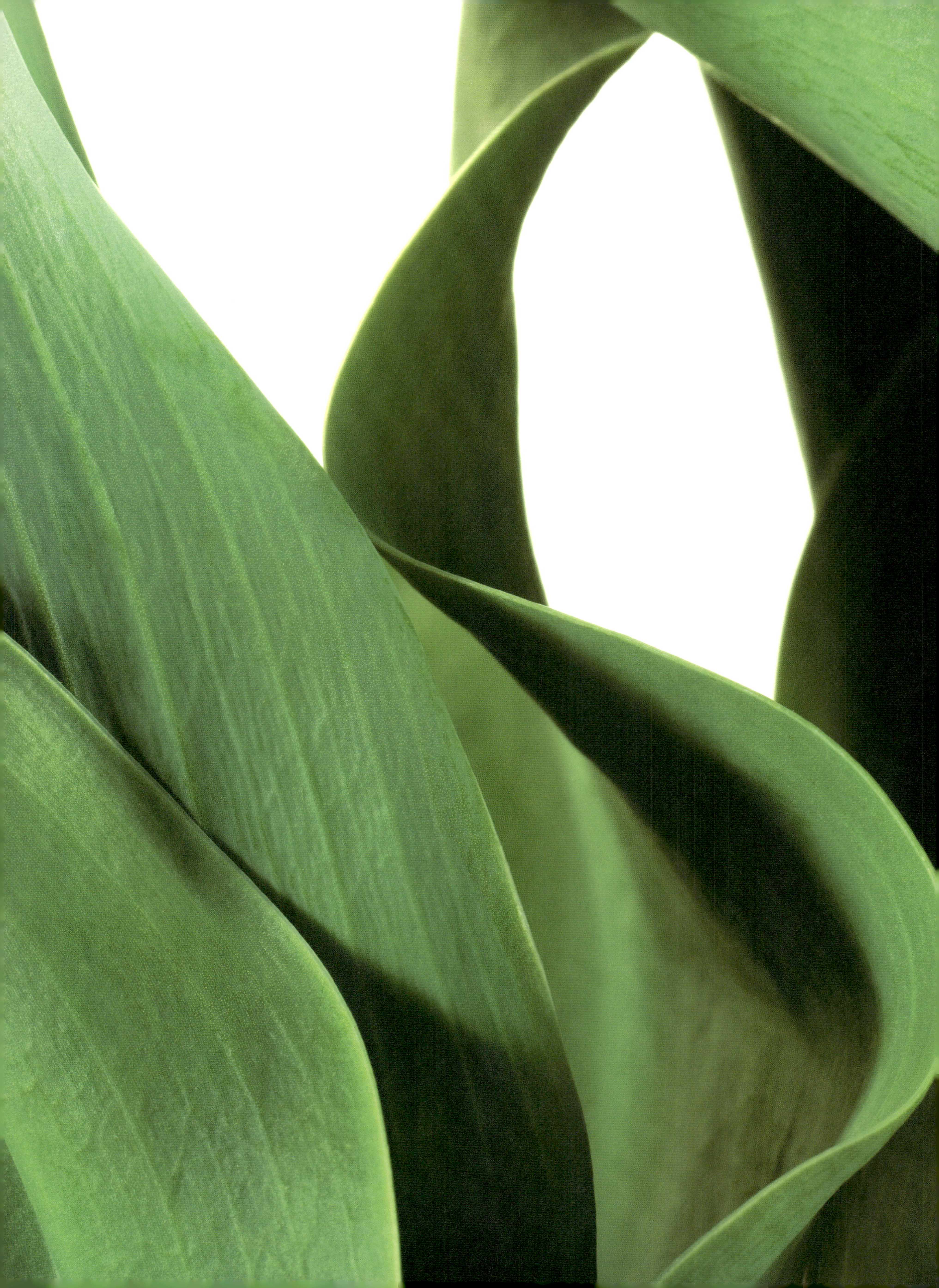

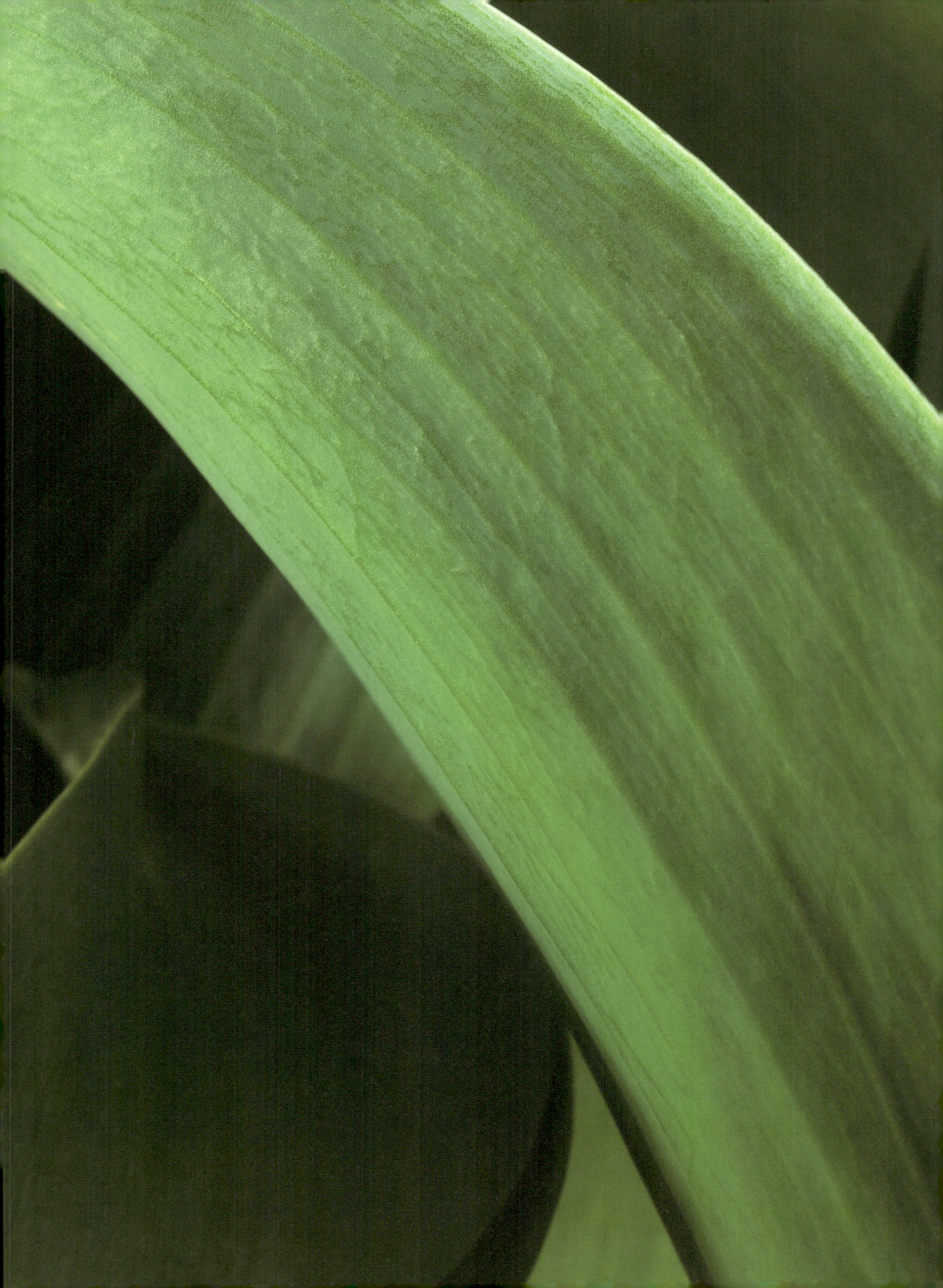

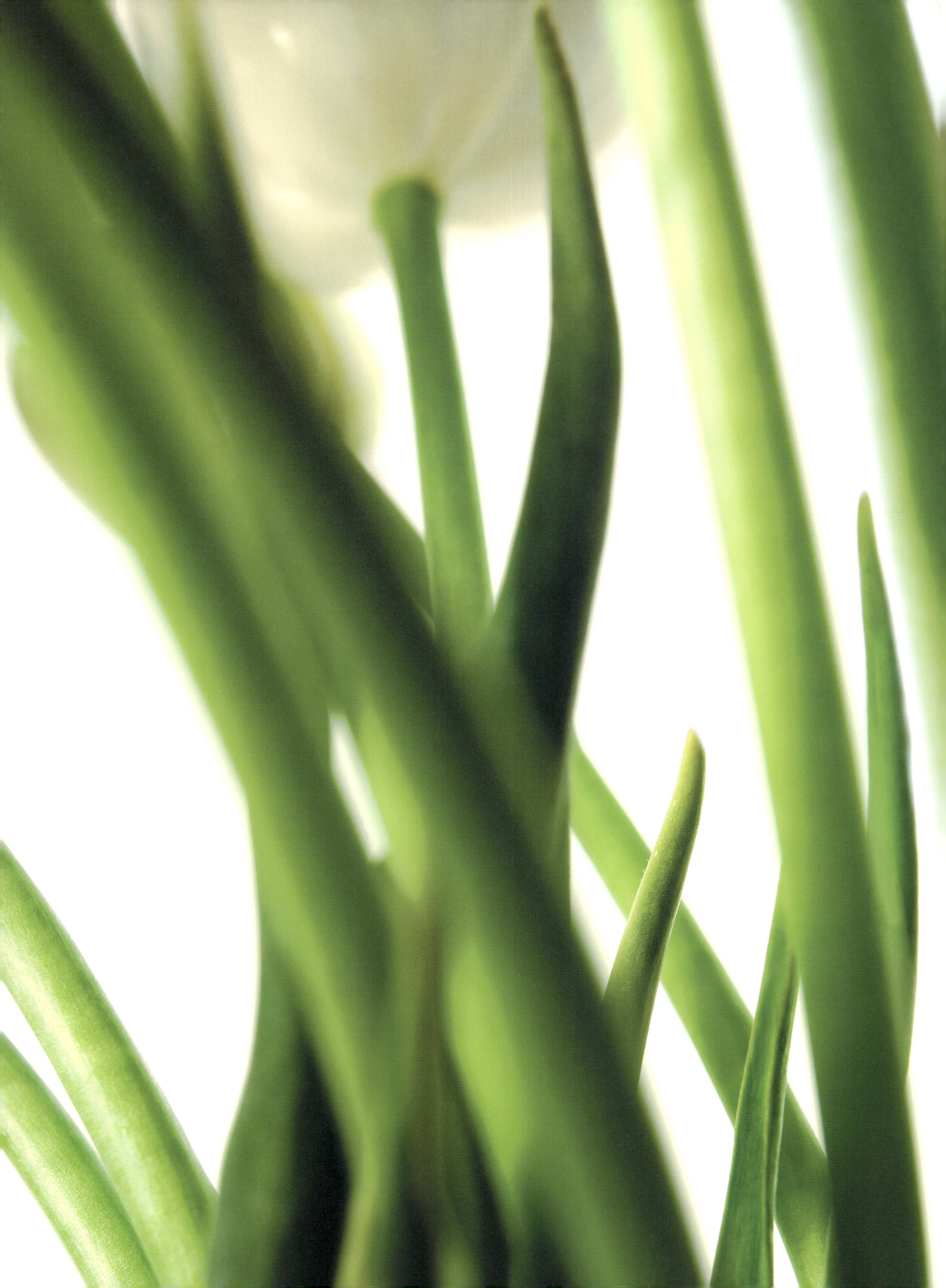

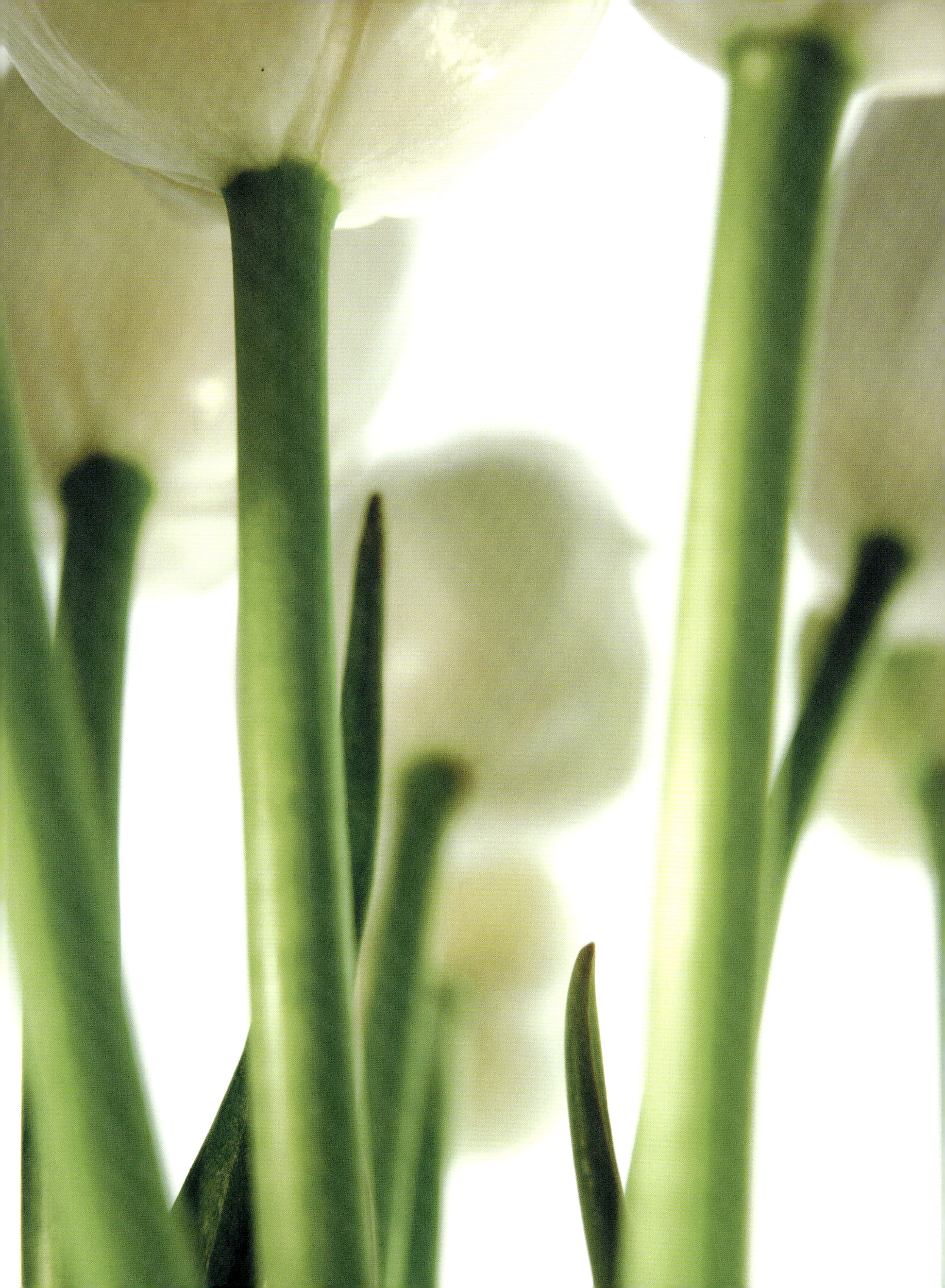

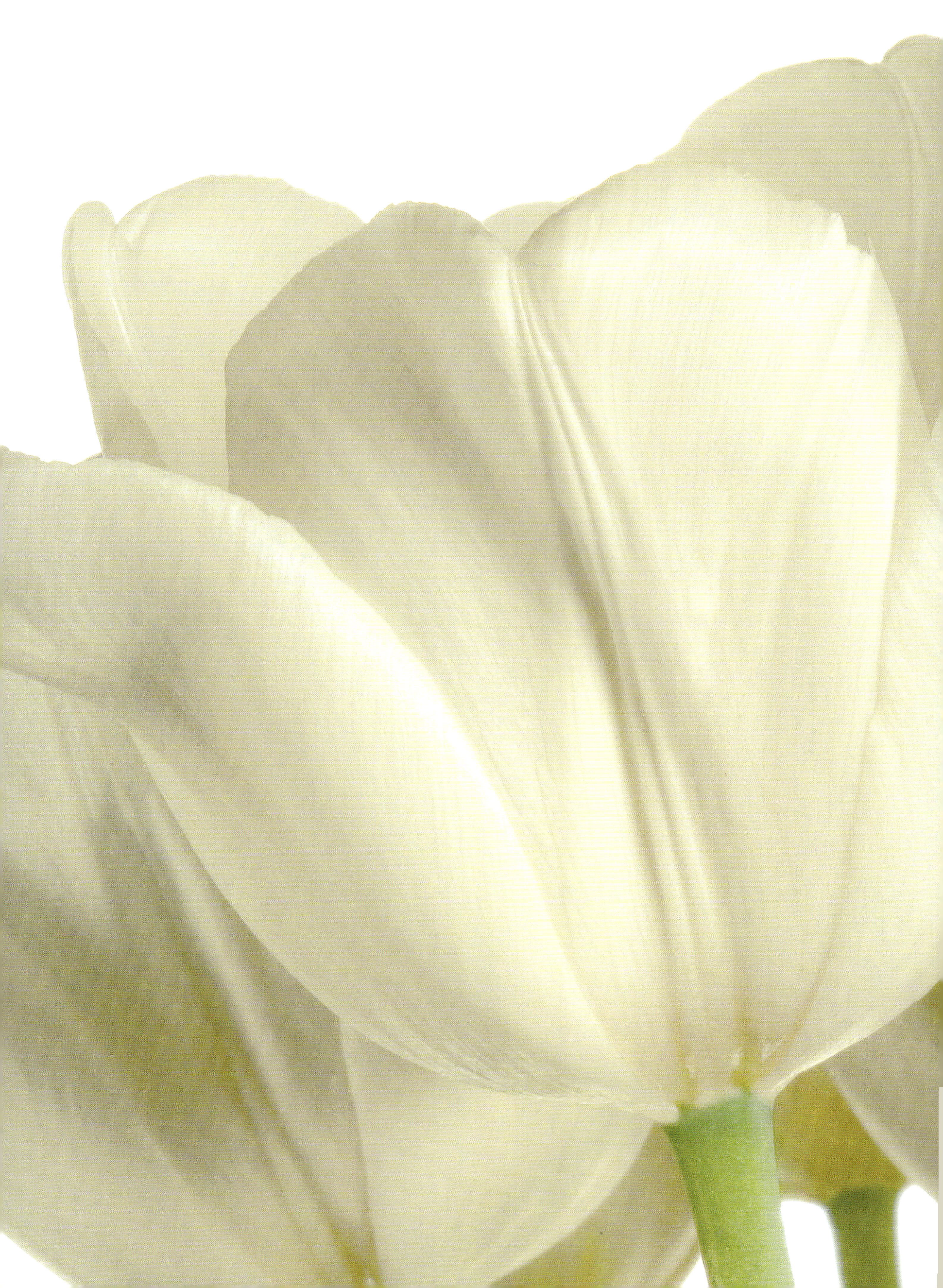

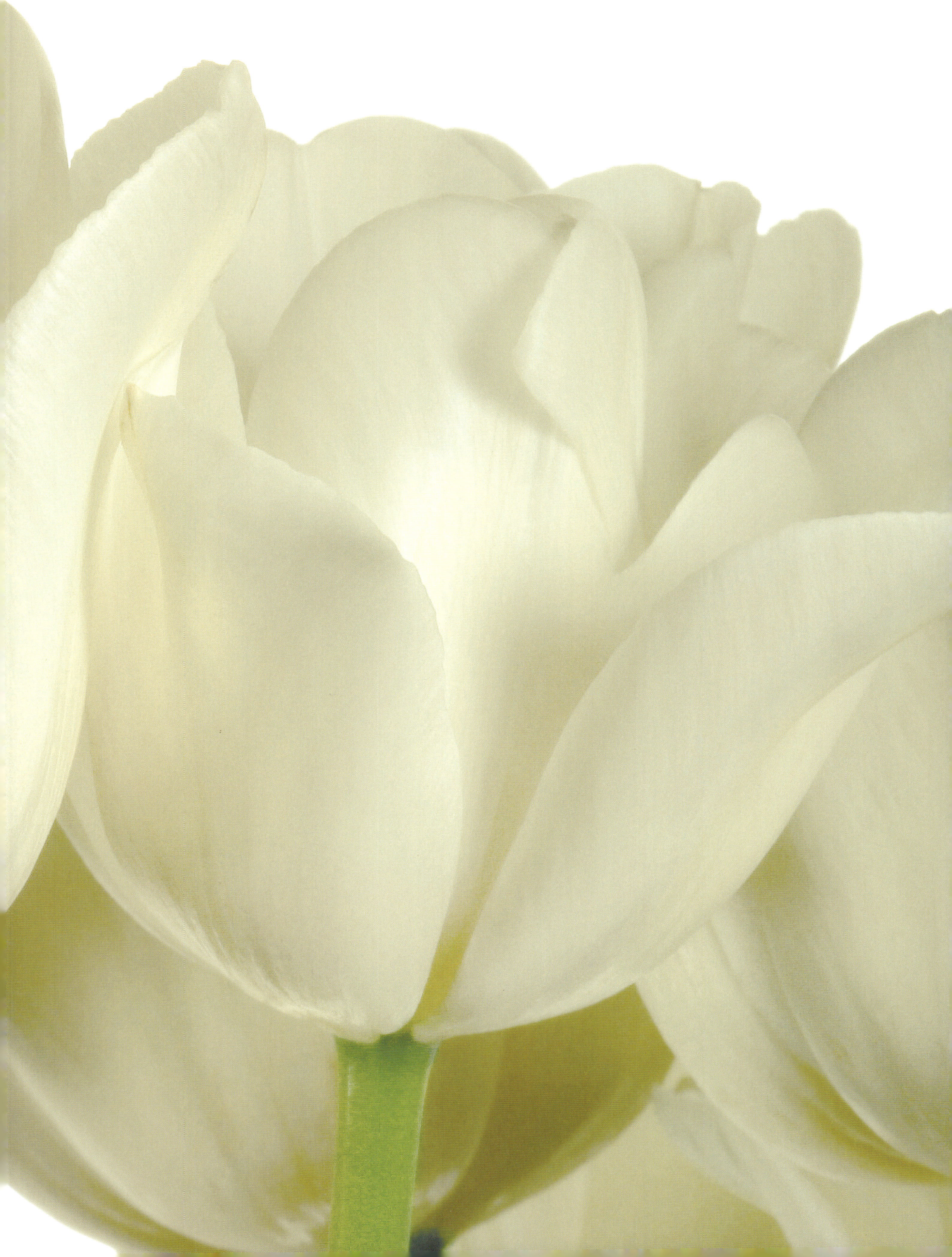

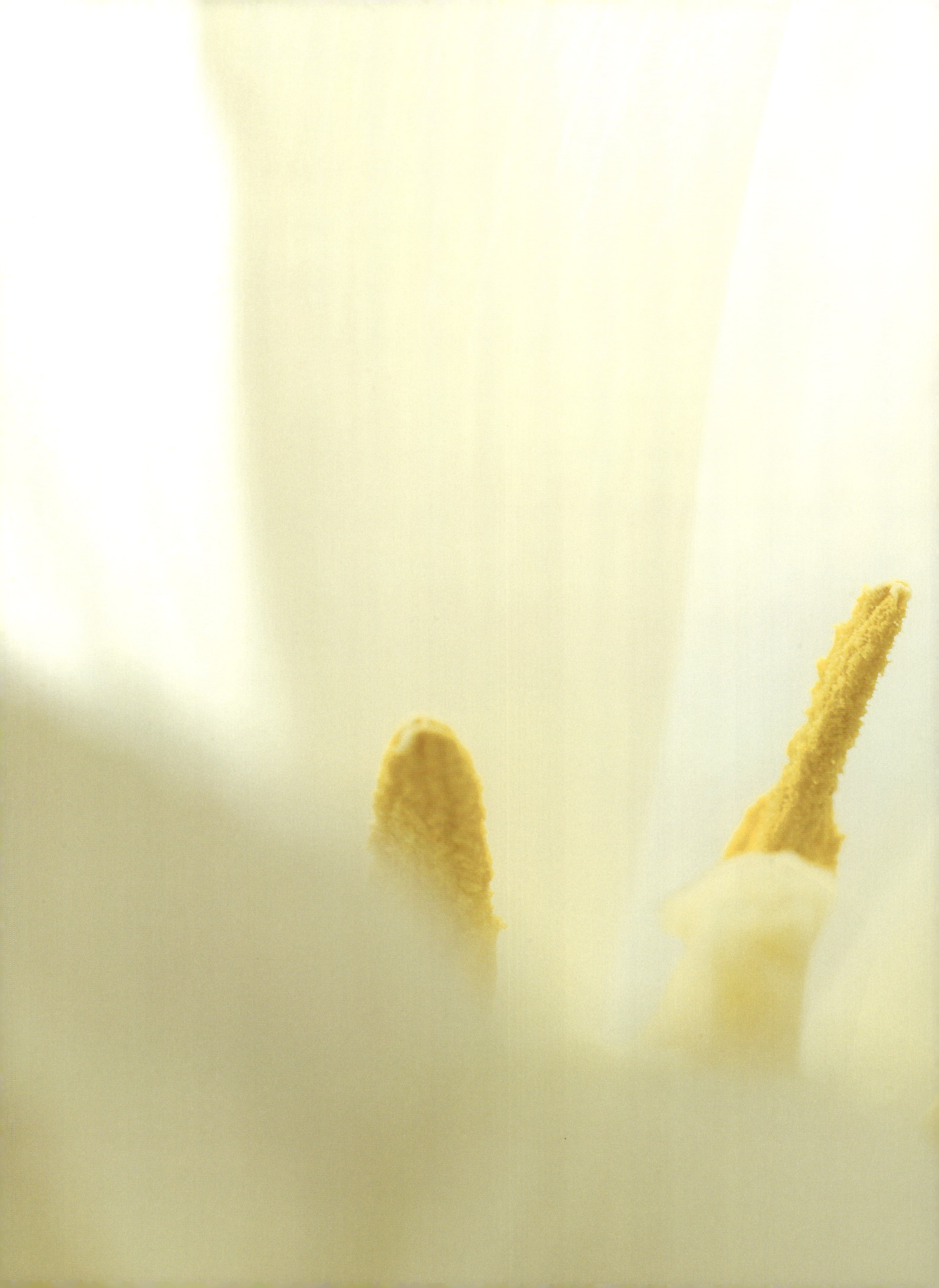

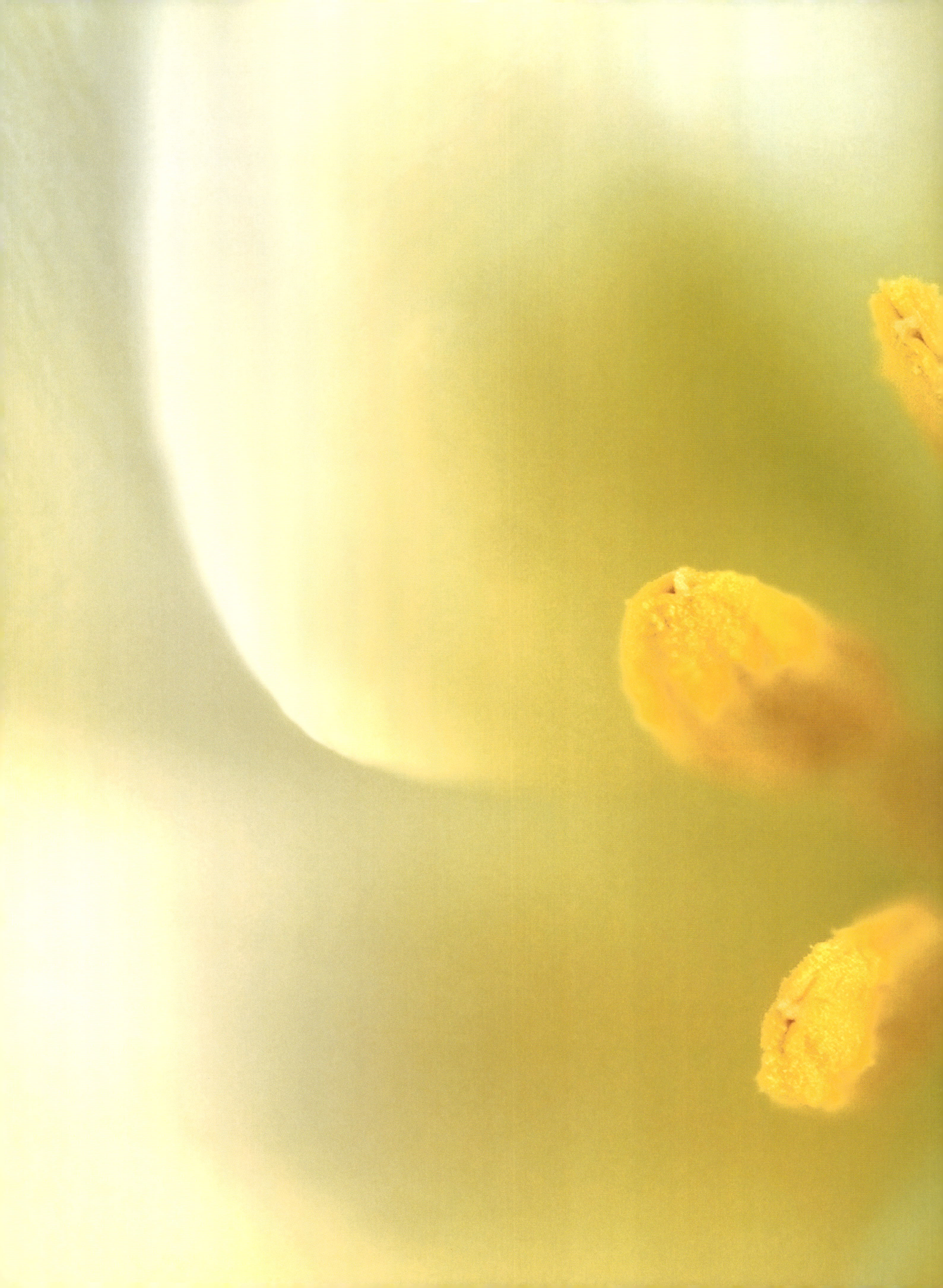

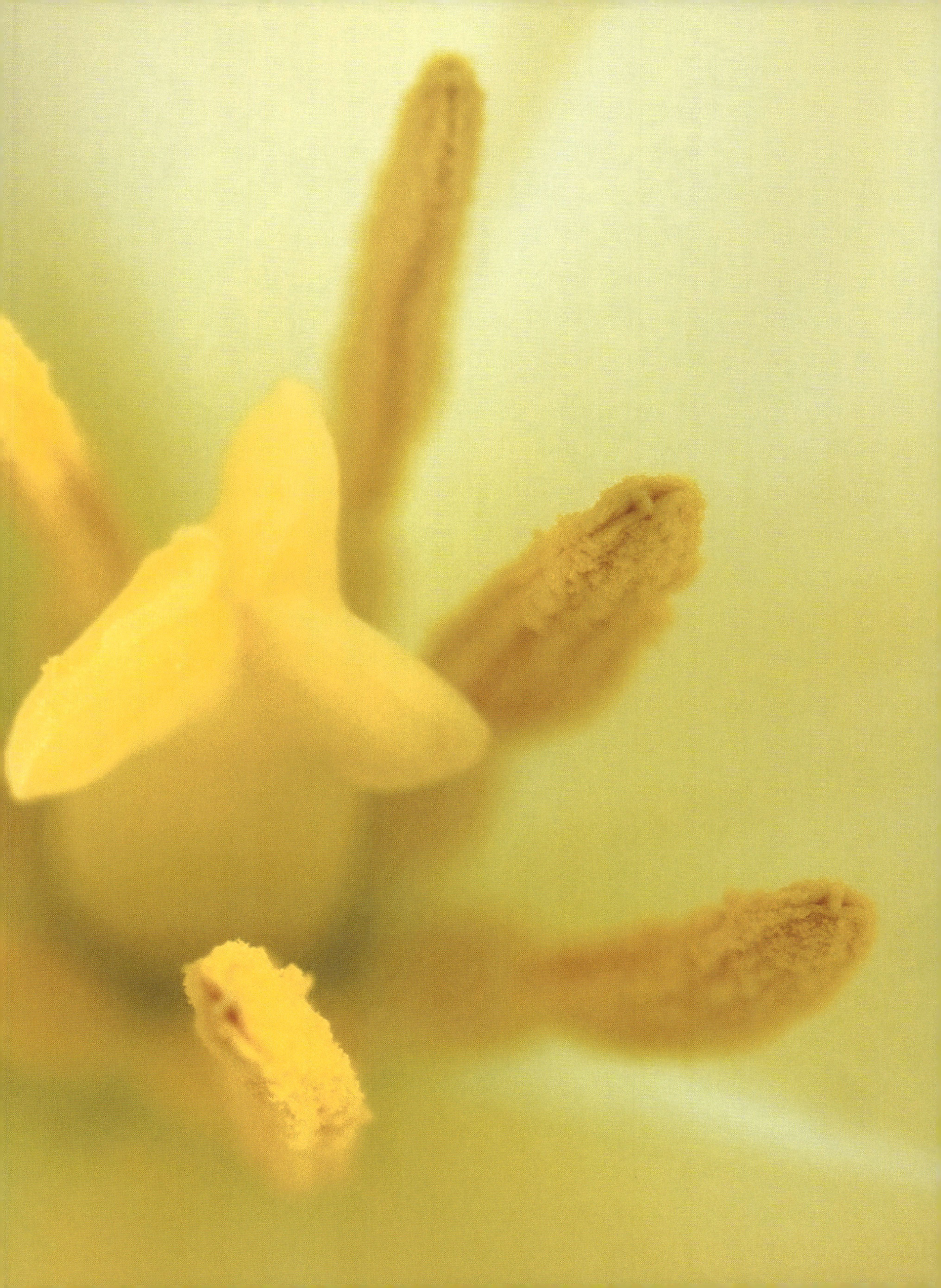

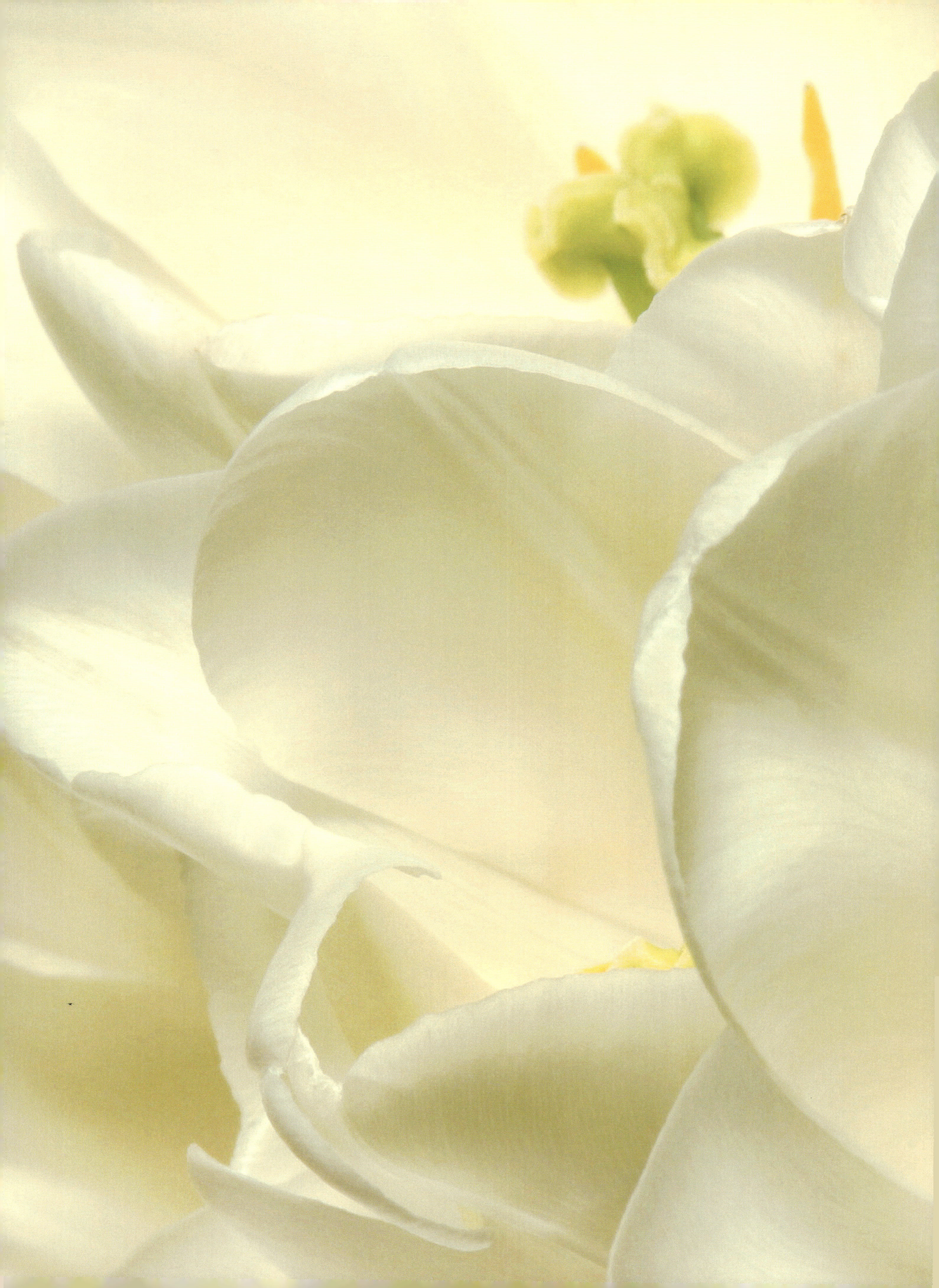

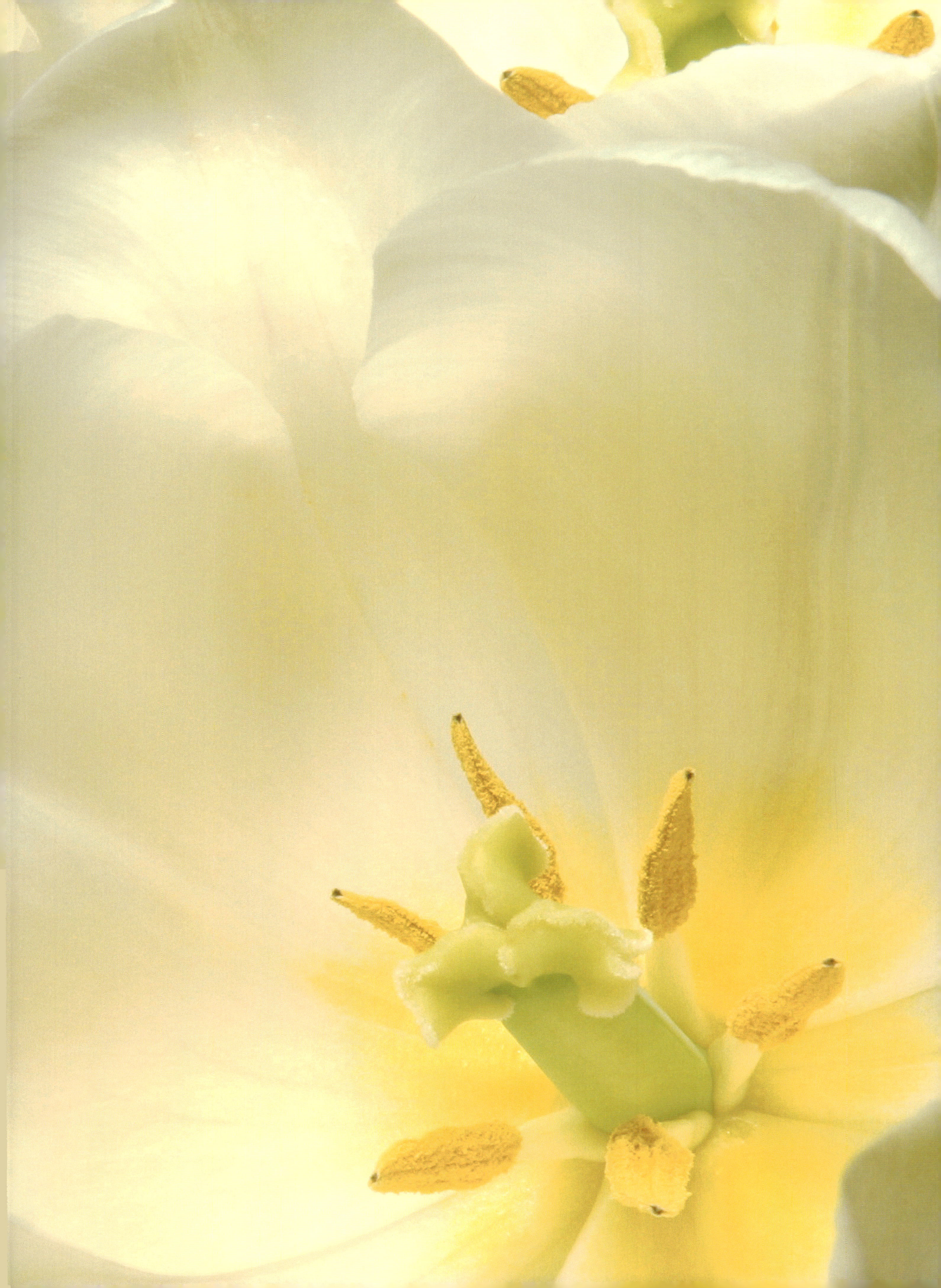

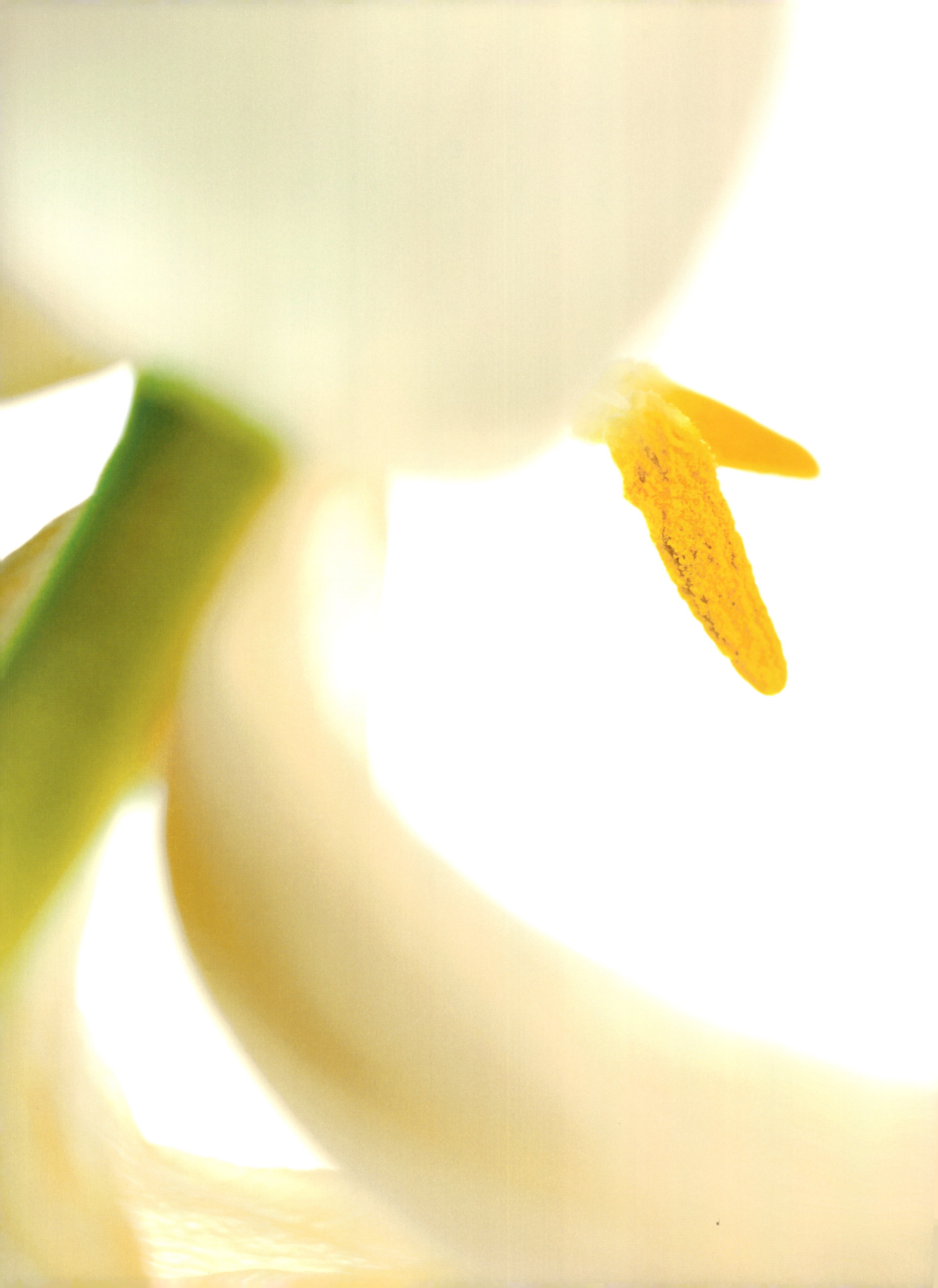

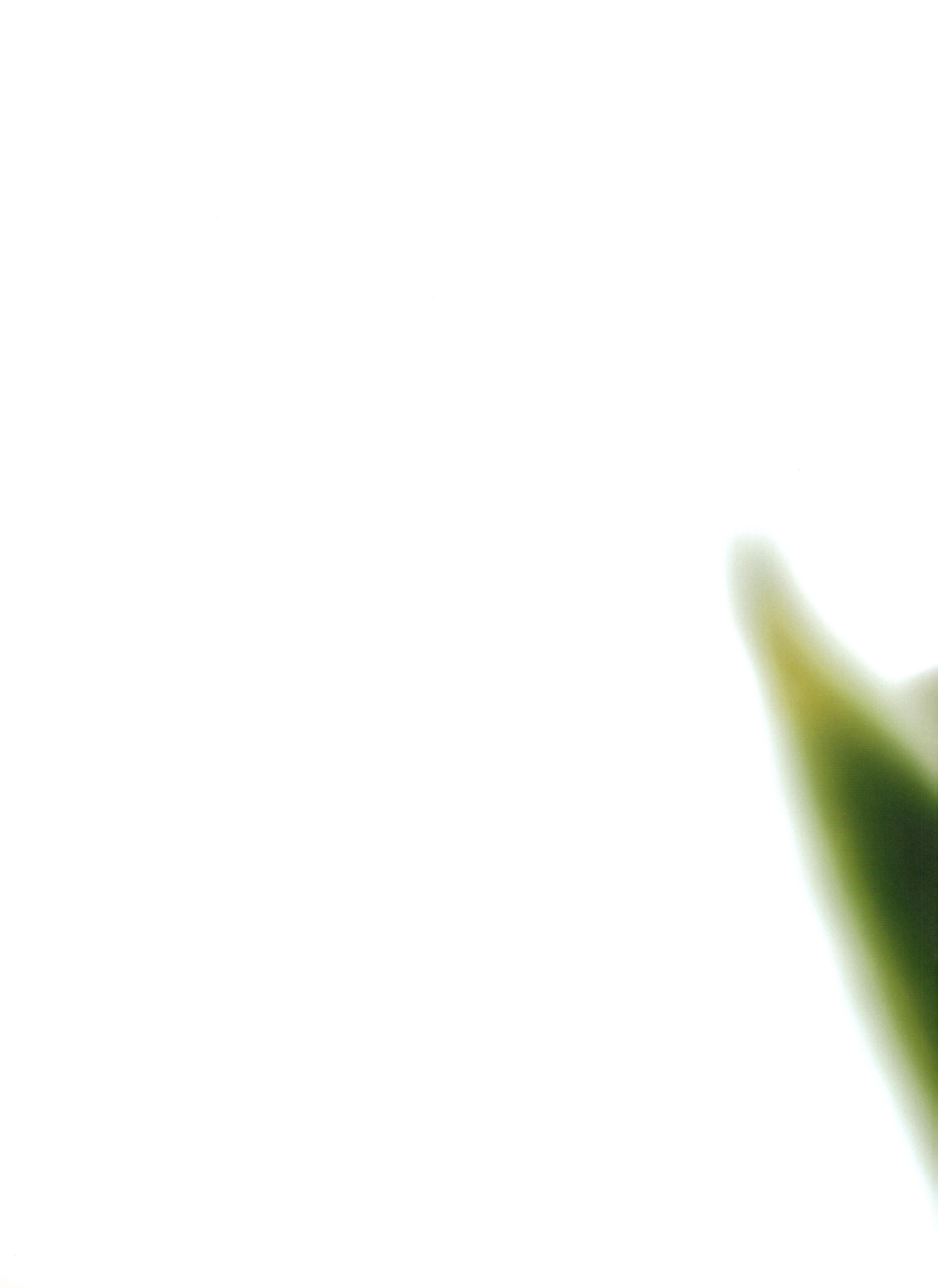

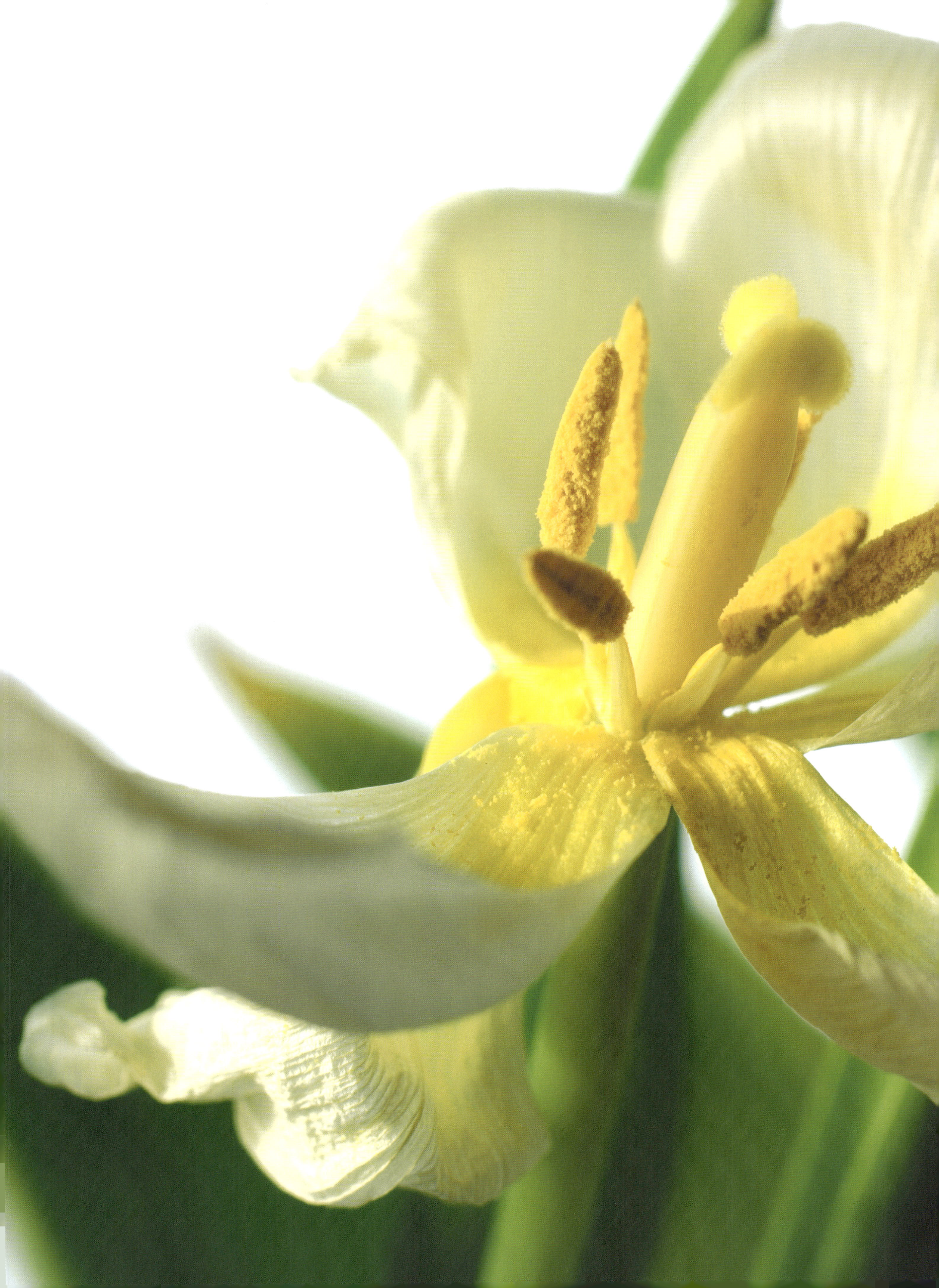

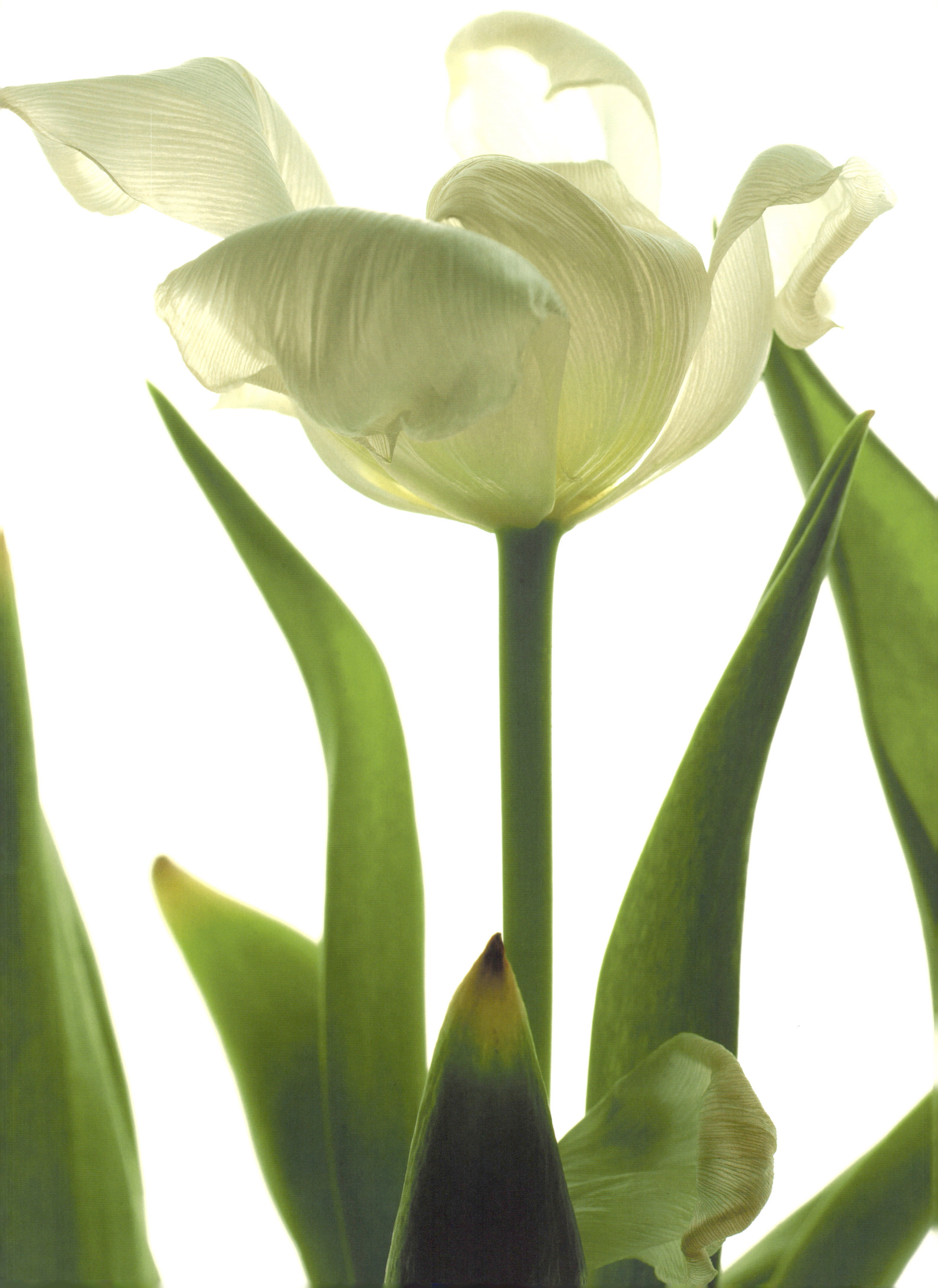

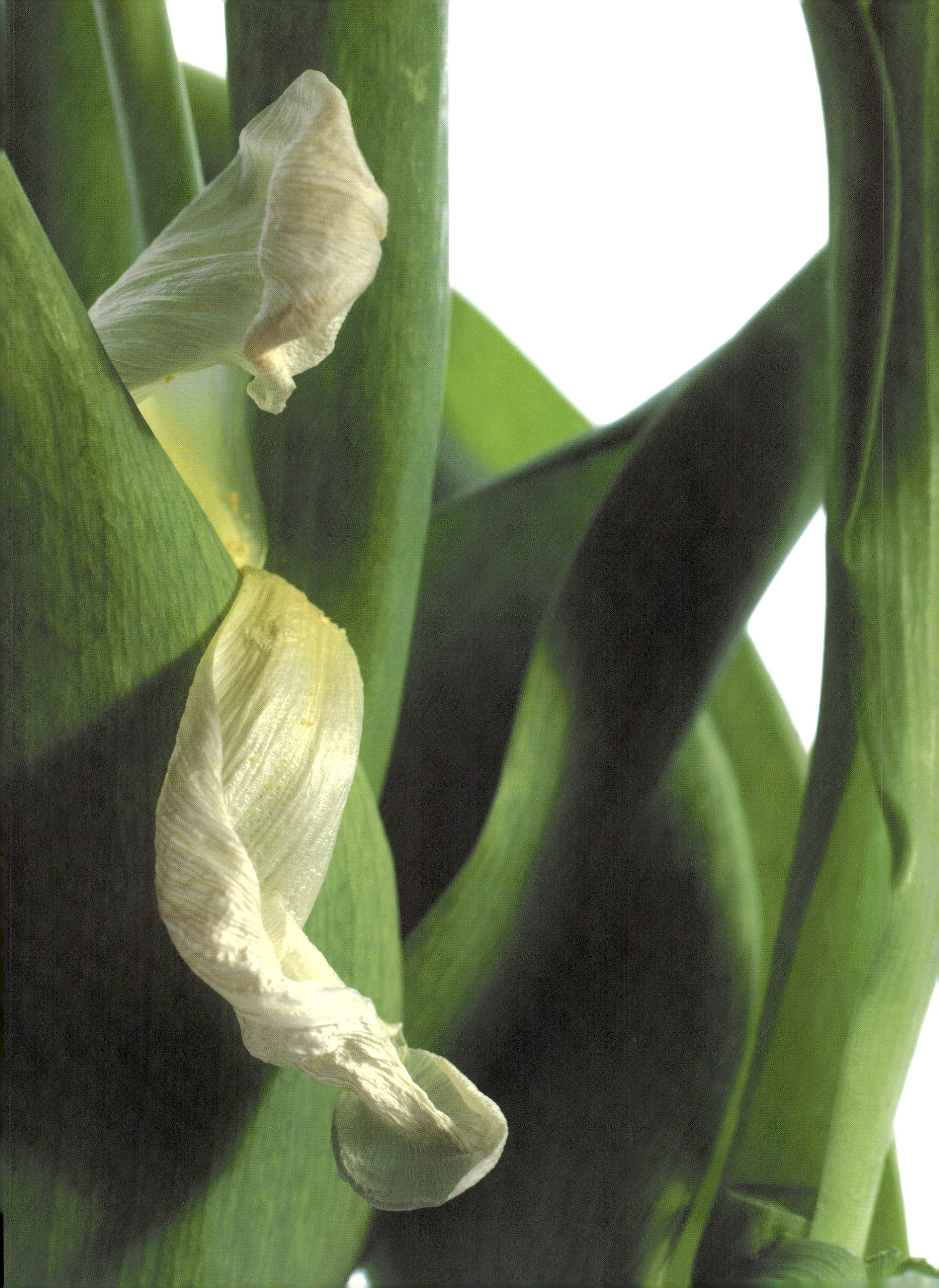

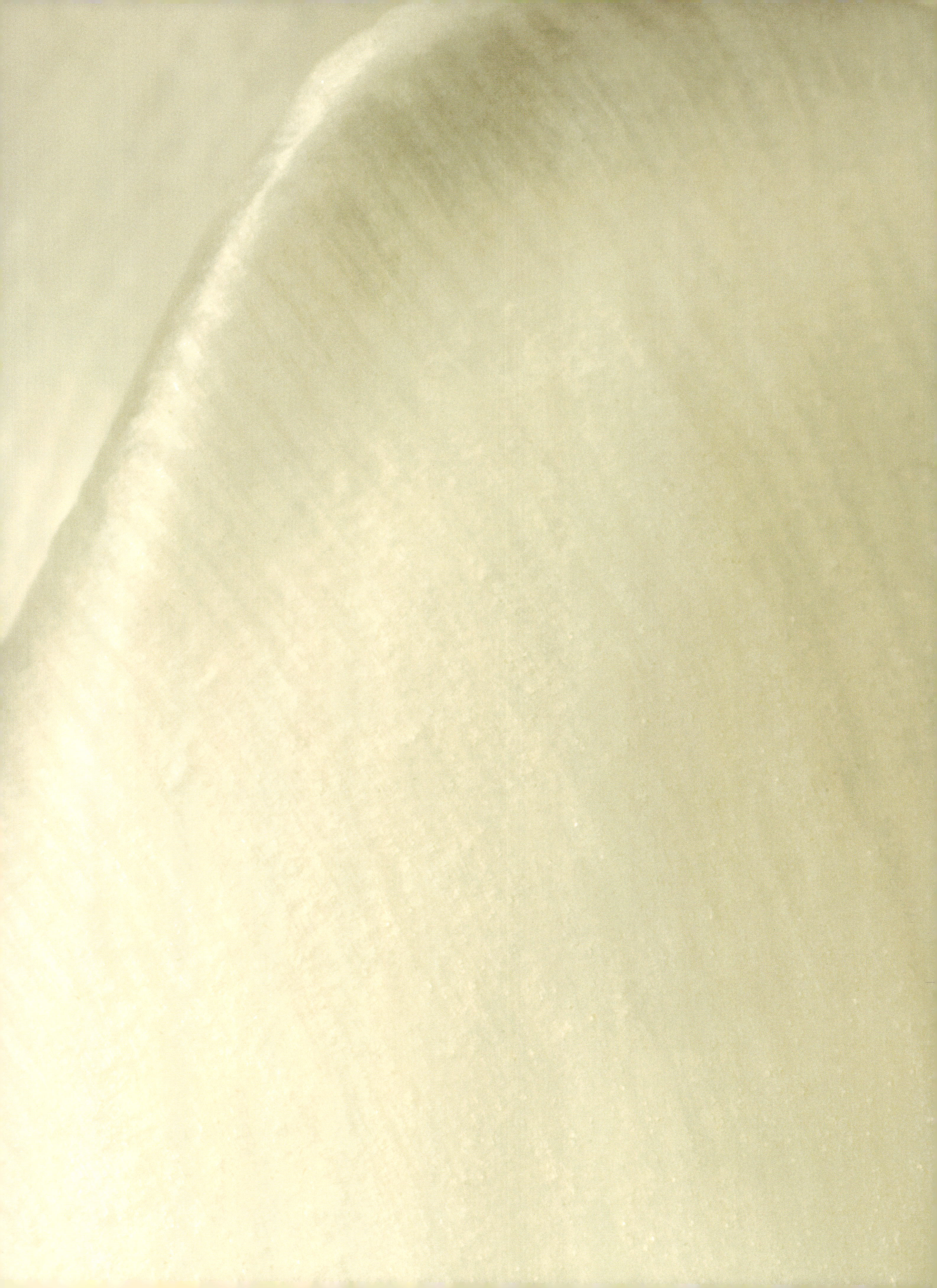

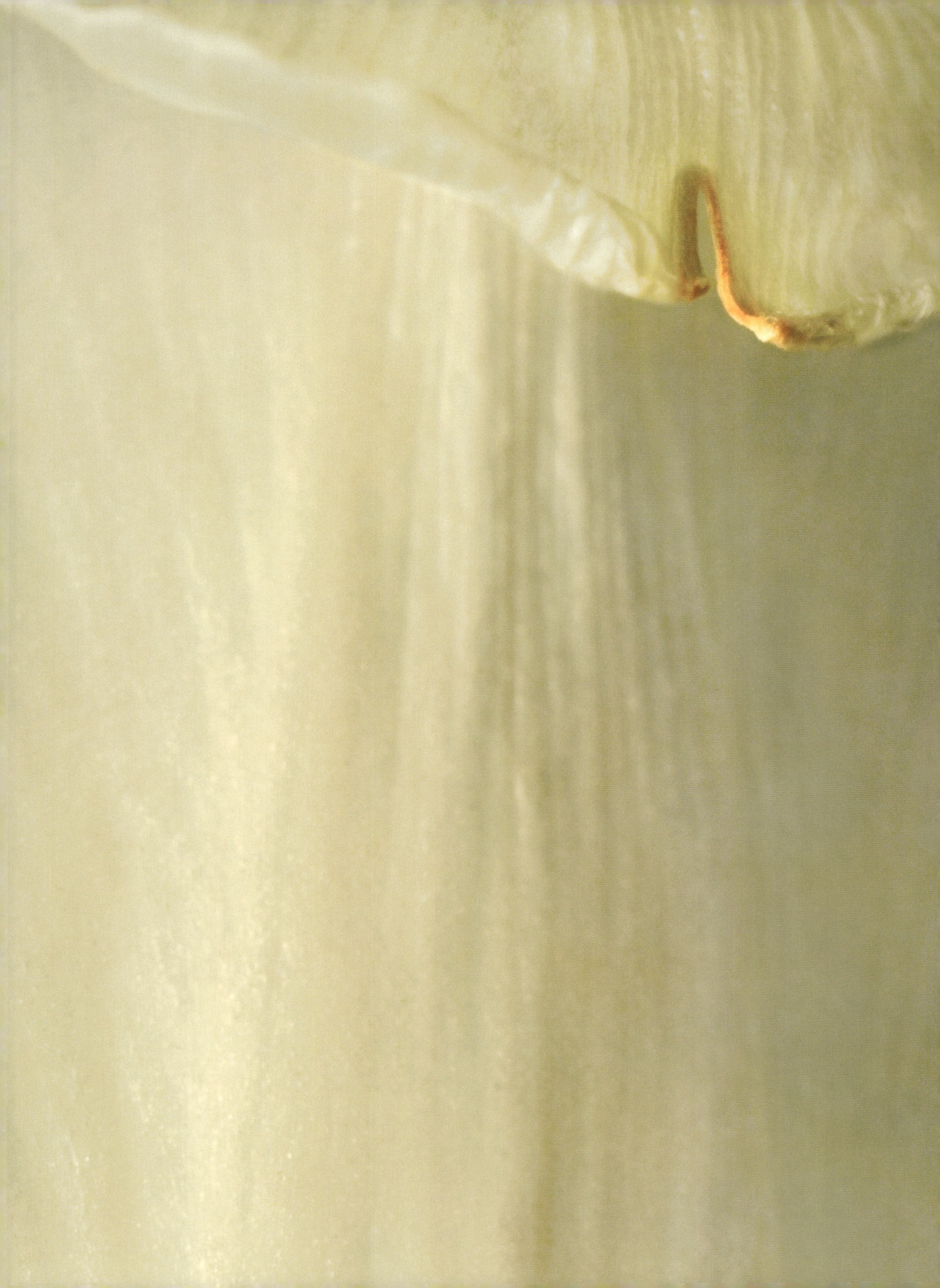

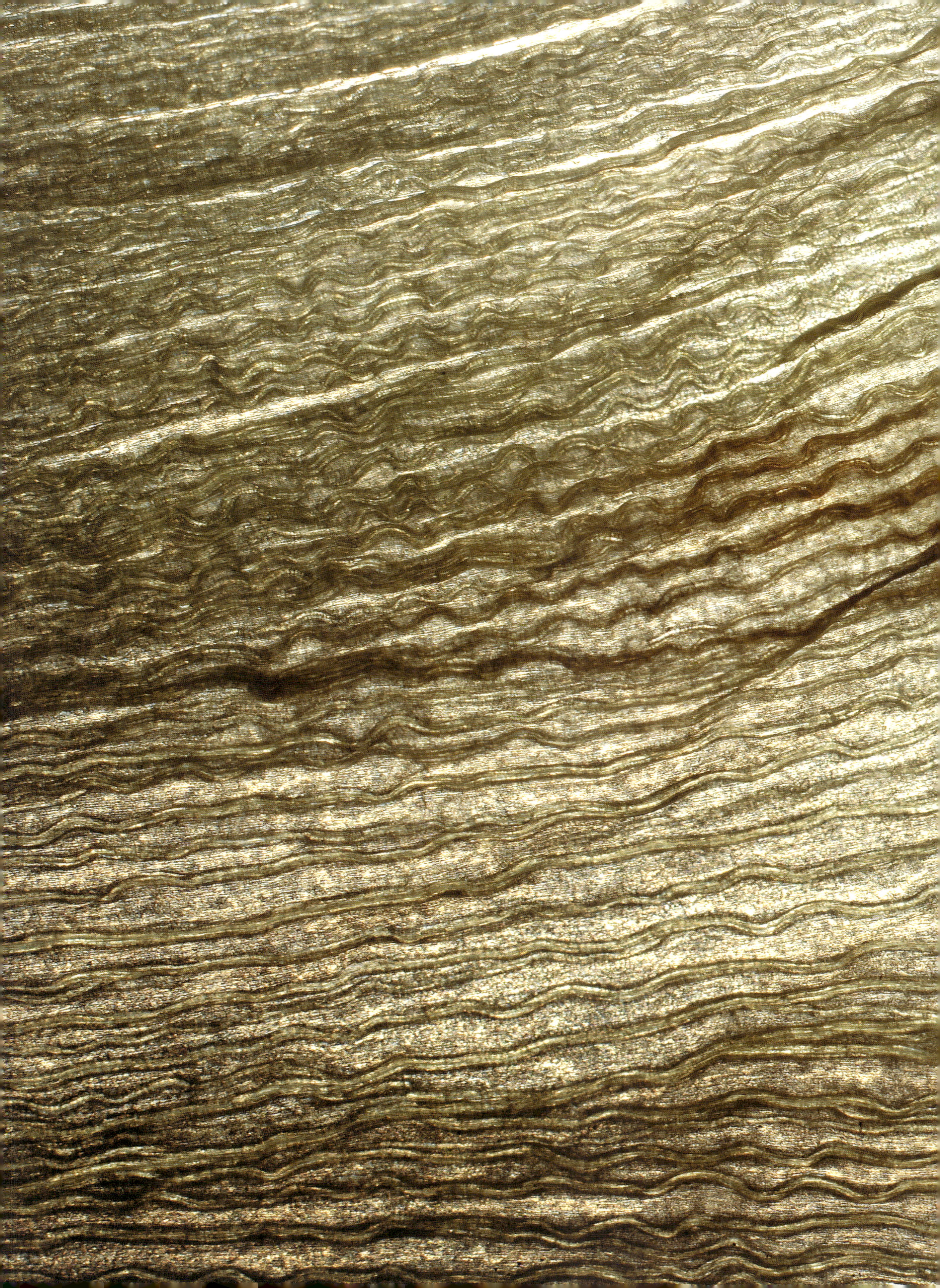

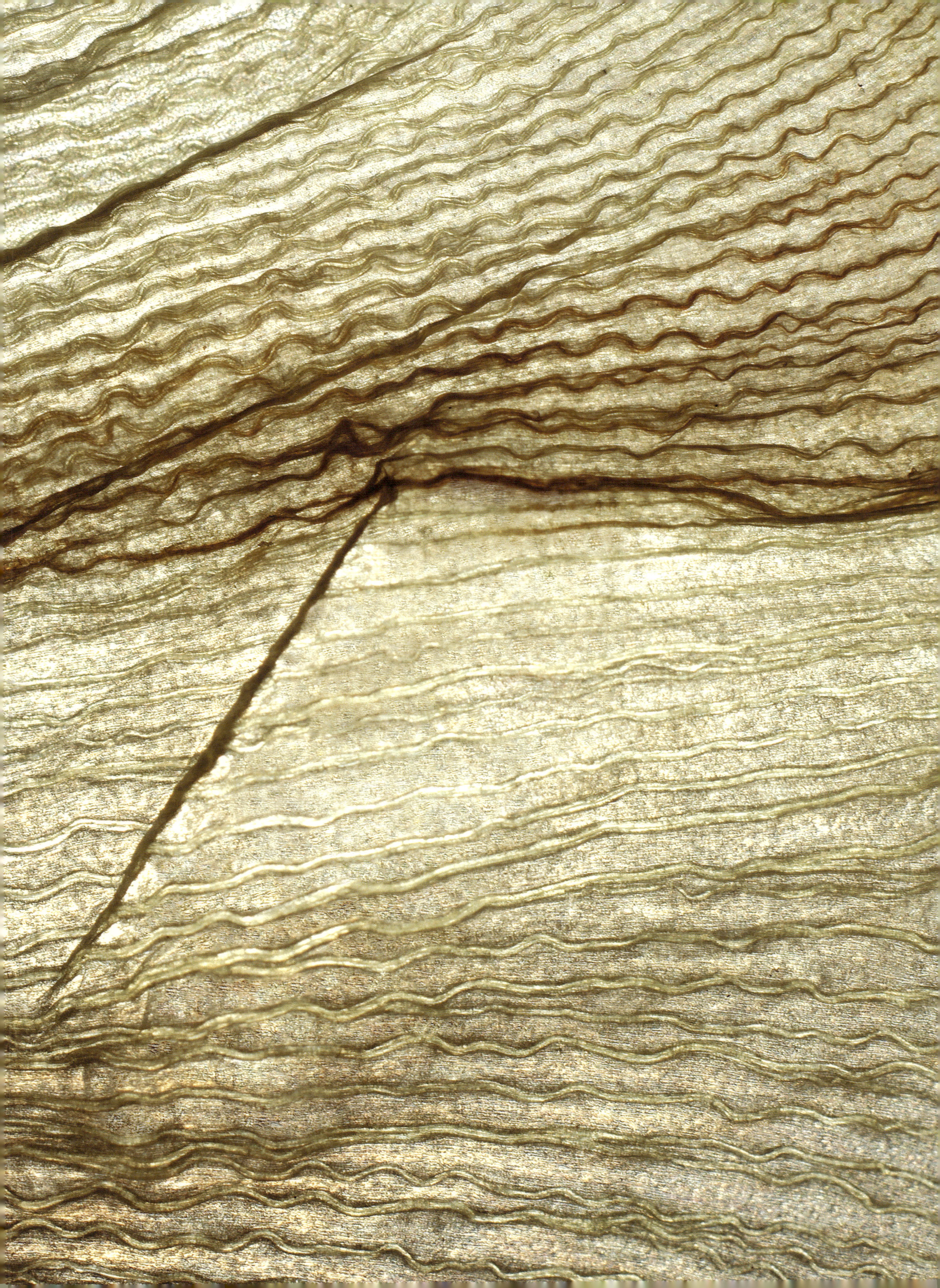

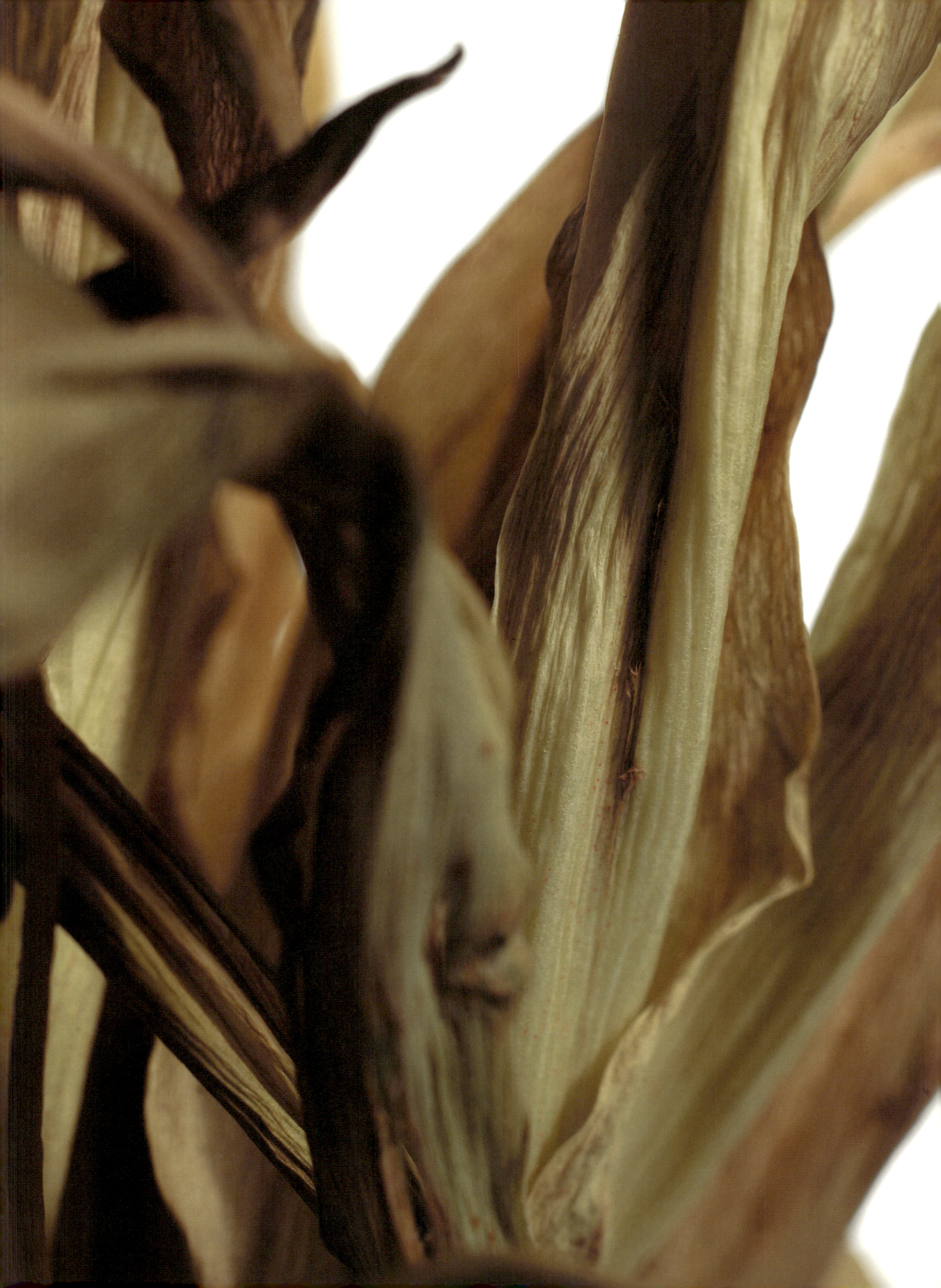

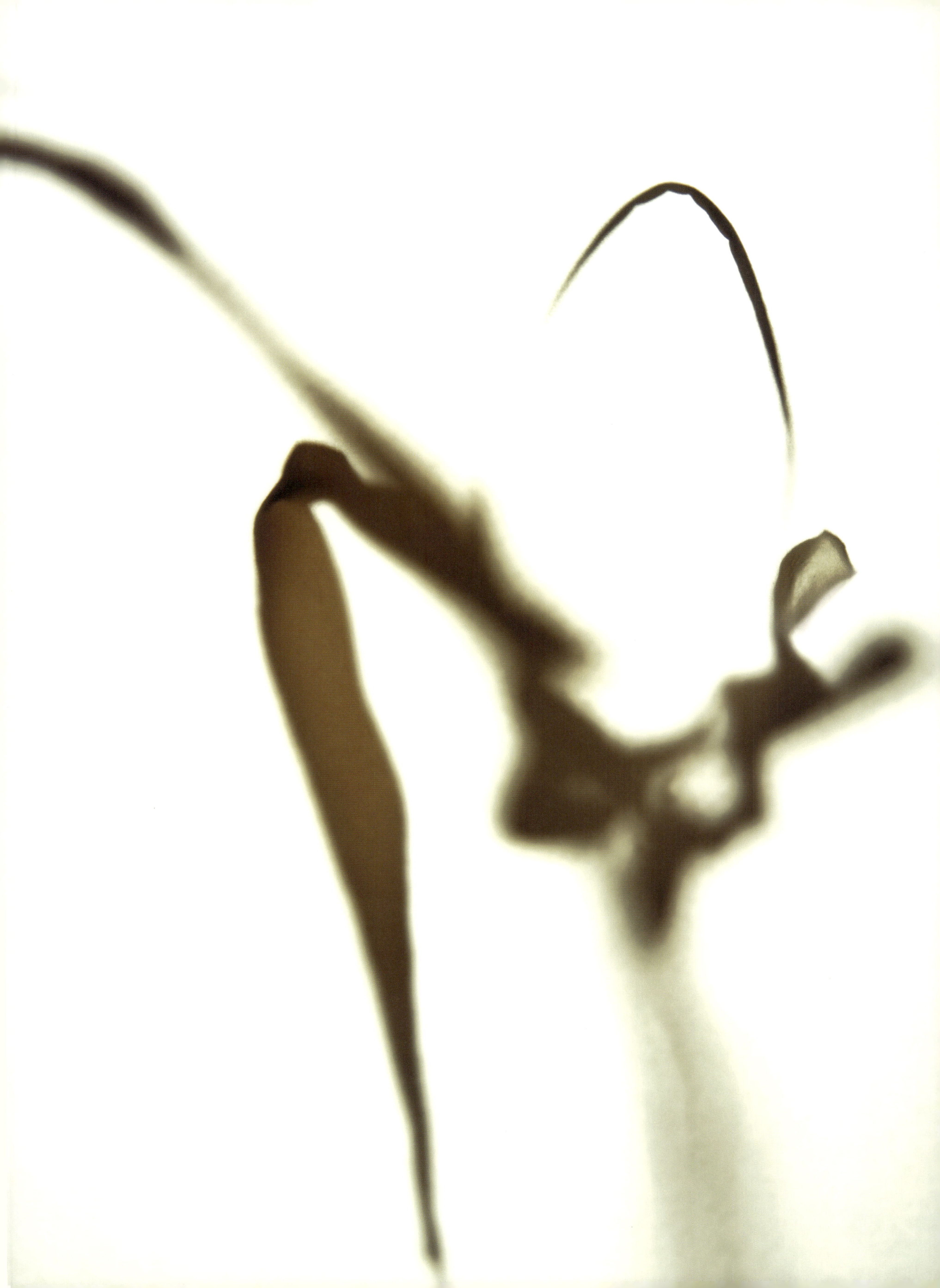

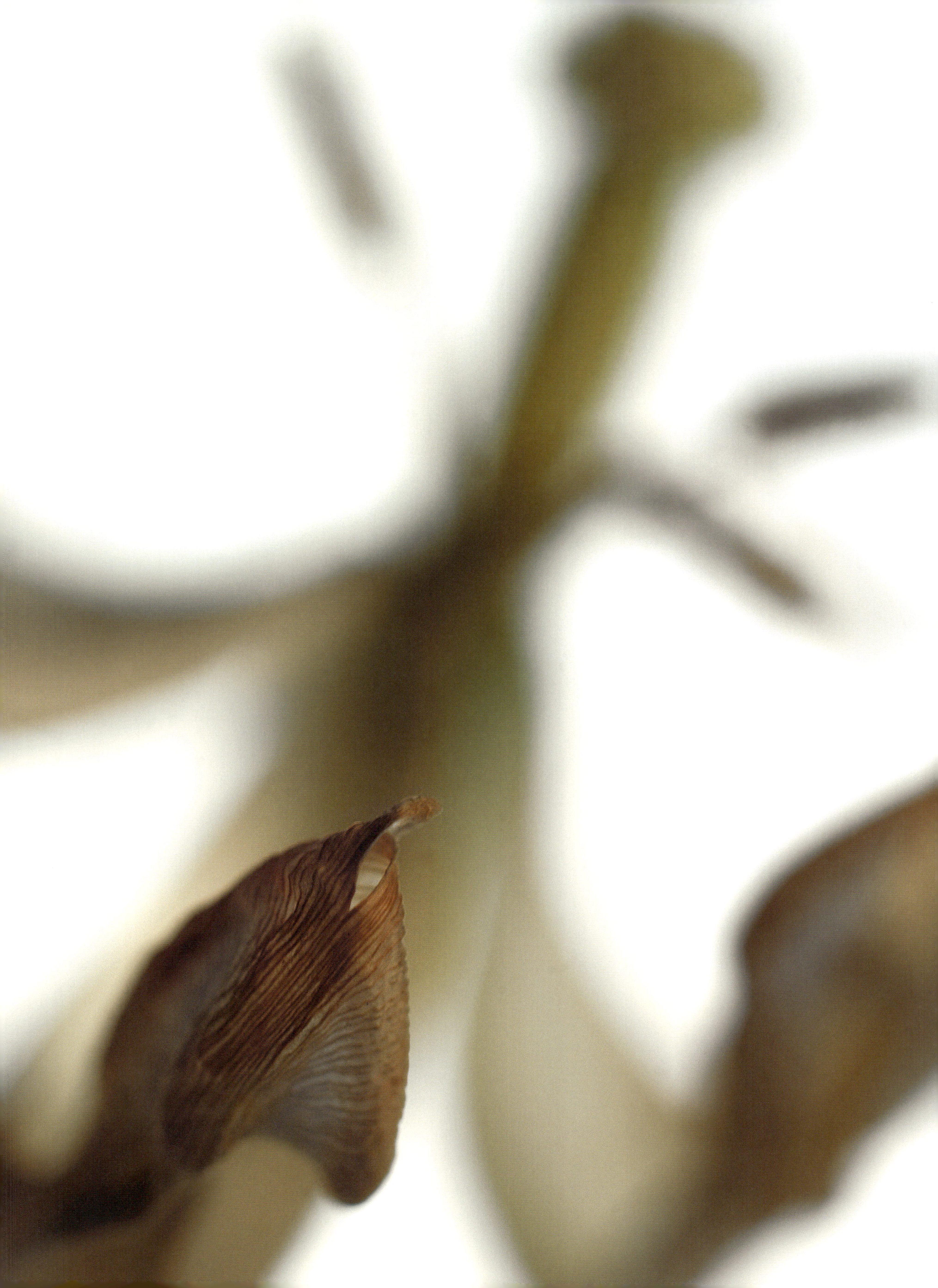

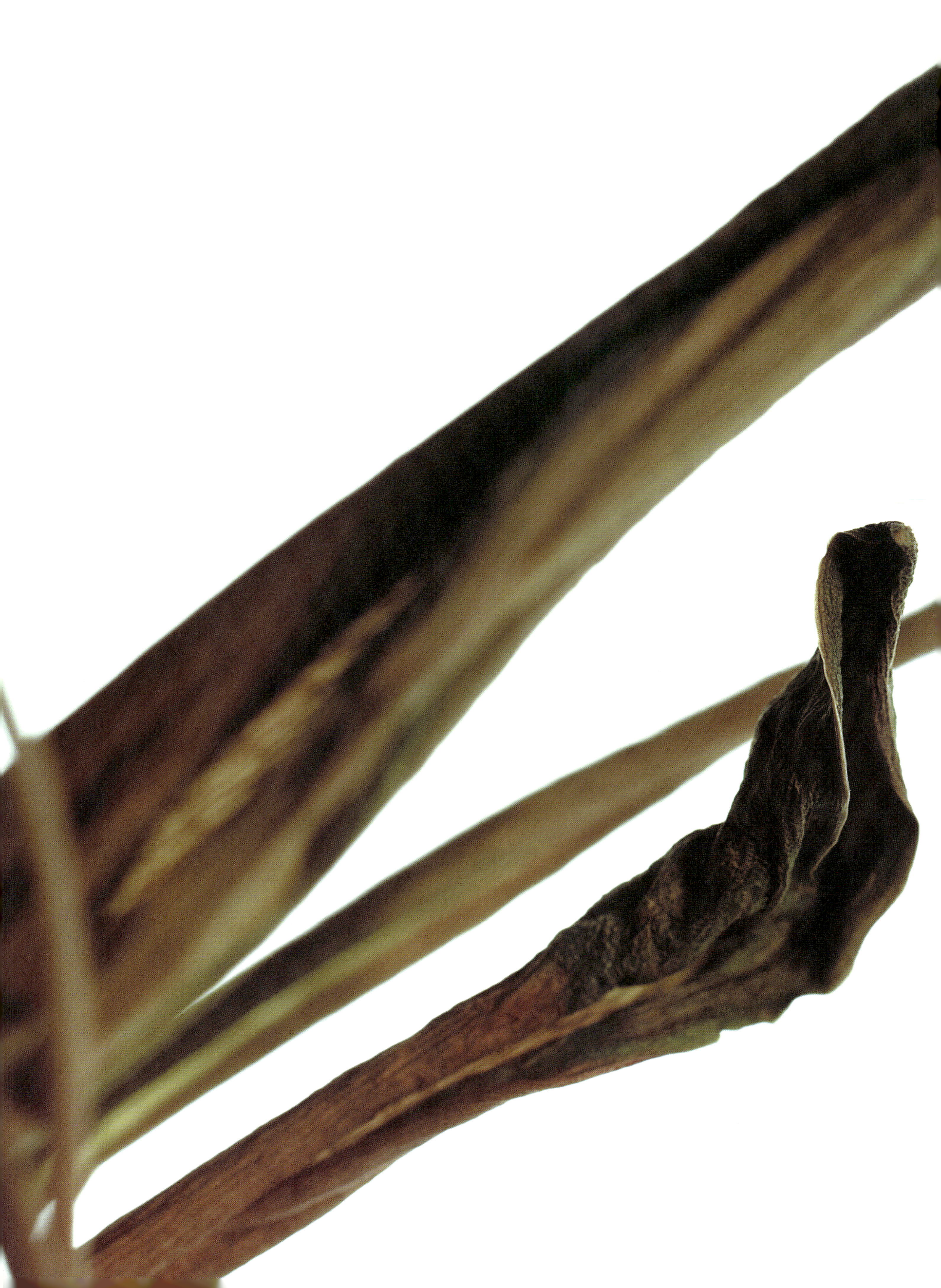

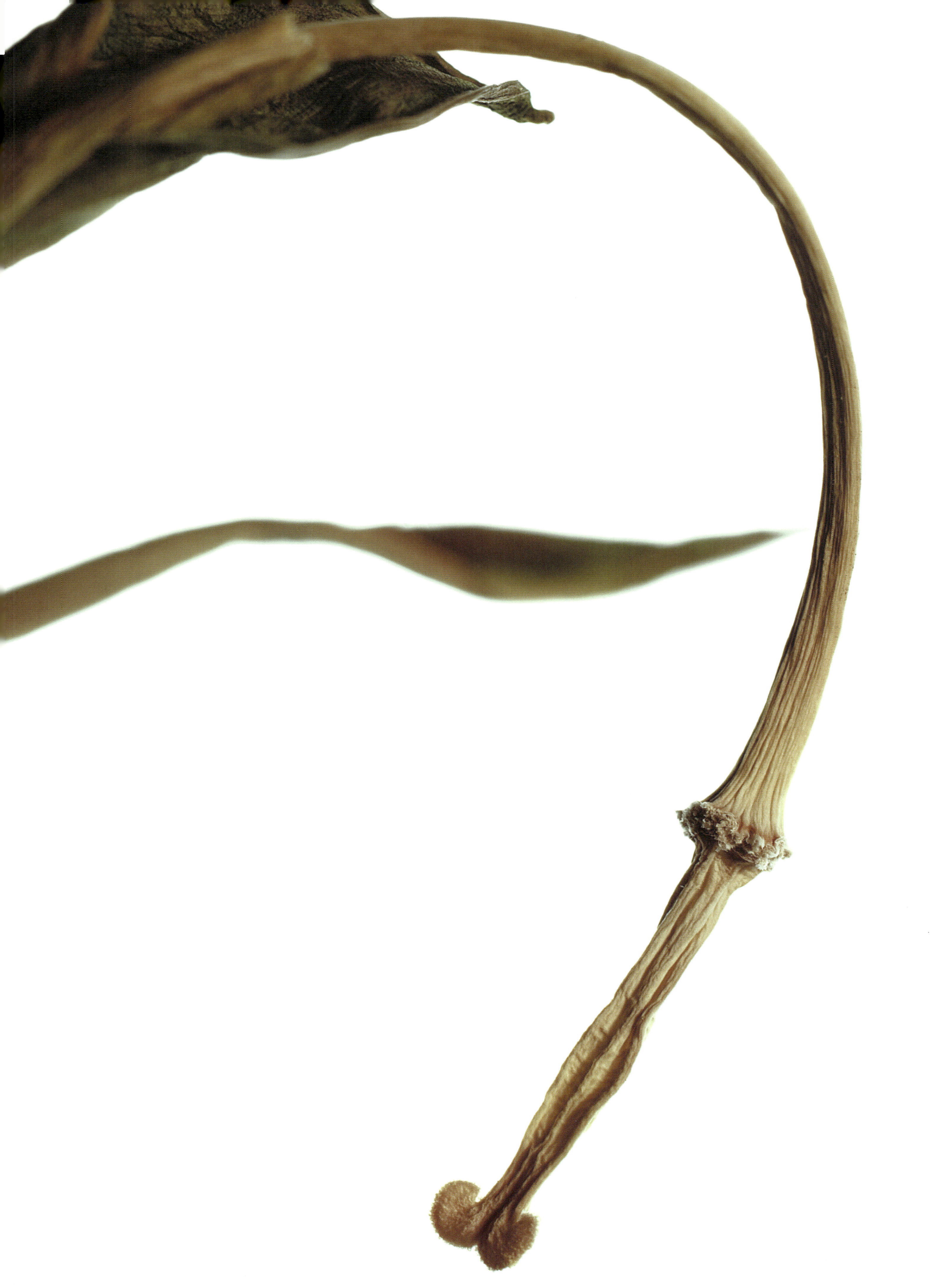

Published by
Verba Volant Ltd.
Atlas House, 4th Floor
1 King Street
London EC2V 8AU

verbavolant
e-mail: info@verbavolant.co.uk
website: www.verbavolant.co.uk

Printed in Italy by Artegrafica S.p.A.

ISBN: 0-9544288-8-9